# TECHNOLOGY AND
# ENVIRONMENT

Jesse H. Ausubel and Hedy E. Sladovich
Editors

National Academy of Engineering

NATIONAL ACADEMY PRESS
Washington, D.C.   1989

National Academy Press • 2101 Constitution Avenue, N.W. • Washington, D.C. 20418

NOTICE: The National Academy of Engineering was established in 1964, under the charter of the National Academy of Sciences, as a parallel organization of outstanding engineers. It is autonomous in its administration and in the selection of its members, sharing with the National Academy of Sciences the responsibility for advising the federal government. The National Academy of Engineering also sponsors engineering programs aimed at meeting national needs, encourages education and research, and recognizes the superior achievement of engineers. Dr. Robert M. White is president of the National Academy of Engineering.

This publication has been reviewed by a group other than the authors according to procedures approved by a National Academy of Engineering report review process. Inclusion of signed work in this publication signifies that it is judged a competent and useful contribution worthy of public consideration, but it does not imply endorsement of conclusions or recommendations by the National Academy of Engineering. The interpretations and conclusions expressed in this volume are those of the authors and are not presented as the views of the council, officers, or staff of the Academy.

Funds for the activity that led to this publication were provided by the Academy's Technology Agenda Program.

**Library of Congress Cataloging-in-Publication Data**

Technology and environment / Jesse H. Ausubel and Hedy E. Sladovich, editors.
      p.    cm.
"National Academy of Engineering."
Bibliography: p.
Includes index.
ISBN 0-309-04075-2
1. Environmental engineering—Technological innovations.
2. Environmental protection.   I. Ausubel, Jesse H.   II. Sladovich, Hedy E.   III. National Academy of Engineering.
TD153. T43    1989                                   89-12674
628—dc20                                                   CIP

*Cover*: Susan Bee, *Landscape with Triangles*, oil on canvas. Courtesy of the artist.

Copyright © 1989 by the National Academy of Sciences

Printed in the United States of America

# Advisory Committee on Technology and Society*

ROBERT M. WHITE, *chairman*, National Academy of Engineering
RUTHERFORD ARIS, University of Minnesota
DAVID P. BILLINGTON, Princeton University
HARVEY BROOKS, Harvard University
DONALD N. FREY, Northwestern University
JOHN H. GIBBONS, Office of Technology Assessment
MARY L. GOOD, Allied-Signal, Inc.
HENRY R. LINDEN, Gas Research Institute
GERALD NADLER, University of Southern California
JAMES BRIAN QUINN, Dartmouth College
WALTER A. ROSENBLITH, Massachusetts Institute of Technology
WALTER G. VINCENTI, Stanford University
ALVIN M. WEINBERG, Oak Ridge Associated Universities

*Membership as of June 1989.

# Preface

It is more evident than ever before that global, regional, and local environmental deterioration is affecting the habitability of the planet and, as a result, societal well-being, the working of the national and global economy, and political interactions among nations. It is also evident that although the management and use of technology have contributed to the present environmental situation, the applications of engineering and technology can provide solutions to many environmental problems. Over the past two decades, concerns have intensified about the decrease of stratospheric ozone, projected climate warming, effects of acid rain, pollution of coastal regions, accumulation of trace pollutants in the biosphere, and increasing difficulties in disposal of nuclear and hazardous wastes. We are now at the point where balancing economic growth and environmental practices makes it ever more urgent to design and apply environmentally innovative engineering.

This volume is an initial contribution in what I hope will be a series of contributions by the National Academy of Engineering (NAE) on the theme of technological innovation and environmental quality. It examines the conceptual framework for thinking about technology and the environment and suggests directions for the education and practice of environmentally responsive engineering. Several major themes emerge from the volume. One is the increasingly complex and important interactions between global markets and the environment. Issues such as chlorofluorocarbon emissions and ozone depletion, fossil fuel use and the greenhouse effect, hazardous waste disposal, and contamination of water resources are no longer merely technical matters but subjects of complex international discussion, negotiation, and bargaining involving industry, governments, environmental groups, and engineering and scientific communities. This volume also points out the need to develop improved intellectual frameworks for analyzing the interactions of technology and environment and to pursue technological opportunities for ameliorating environmental conditions. Finally, the volume suggests the need for greater participation of the technology community in emerging international programs on global environmental change.

Naturally, any such work is partial in its coverage. The links between environmental engineering and economics must be explored further. The focus here is primarily on the United States and other industrialized nations, and vital questions of technology and environment for developing countries remain to be examined in greater depth. Other topics that come to mind, which would complement this volume, include the role of engineering in relation to biota and ecosystem conservation, and the relationship of an aging society to environmental quality as affected by technology. A concurrently published book from the NAE, *Energy: Production, Consumption, and Consequences* (J. Helm, ed.), goes into more detail than is appropriate here on topics such as energy efficiency and conservation.

This volume is the fruit of two years' effort, including a workshop at Woods Hole, Massachusetts, in August 1988 and a symposium, "Technological Opportunities and Environmental Change," in Washington, D.C., in September 1988 during the annual meeting of the NAE. The workshop and planning effort were chaired by Robert Frosch. The concept for the effort may be traced back to correspondence and discussions involving Jesse Ausubel, William Clark, Robert Herman, William Nierenberg, and Stephen Schneider. I would especially like to thank these people, as well as Floyd Culler, who chaired the symposium. Among others who provided valuable ideas and contributions were Alvin Alm, John Brown, David Chittick, Richard Conway, James Coulter, Joel Darmstater, John Farrington, John Helm, Donald Hornig, Thomas Larson, Alfred Lindsey, June Lindstedt-Siva, Joseph Ling, Gregg Marland, William Nordhaus, Frank Parker, Ellen Silbergeld, Lee Thomas, Ian Torrens, and Edwin Zebroski. We are indebted to the authors for their excellent chapters and to an editorial group consisting of Jesse Ausubel, Dale Langford, Hedy Sladovich, and Bette Janson. Nancy McDermott and James Porter provided additional support.

The project was carried out under the overall auspices of the NAE's Technology and Society Advisory Committee and the Program Office of the NAE, directed by Bruce Guile.

As one of the reviewers of the manuscript noted, its essence is captured in the opening essay, which suggests that "the intersection of technology and environment in a sense has been the blind spot in our system of knowledge, and this gap is at the root of today's environmental crisis. . . . Environmental engineering, recognizing our own nature as part of nature and our technology as in nature, can help bridge the dangerous compartmentalization of knowledge and professions that appears to be placing modern life in jeopardy."

ROBERT M. WHITE
President
National Academy of Engineering

# Contents

Technology and Environment: An Overview ........................... 1
  Jesse H. Ausubel, Robert A. Frosch, and Robert Herman

### 1. FRAMEWORKS FOR ANALYSIS

Industrial Metabolism ............................................. 23
  Robert U. Ayres

Dematerialization ................................................. 50
  Robert Herman, Siamak A. Ardekani, and Jesse H. Ausubel

Regularities in Technological Development: An Environmental View ... 70
  Jesse H. Ausubel

### 2. THE PROMISE OF TECHNOLOGICAL SOLUTIONS

Meeting the Near-Term Challenge for Power Plants ................... 95
  Richard E. Balzhiser

Advanced Fossil Fuel Systems and Beyond ........................... 114
  Thomas H. Lee

Protecting the Ozone Layer: A Perspective from Industry ............ 137
  Joseph P. Glas

## 3. SOCIAL AND INSTITUTIONAL ASPECTS

The Rise and Fall of Environmental Expertise ........................ 159
    *Victoria J. Tschinkel*

Environmental Issues: Implications for Engineering Design
and Education ..................................................... 167
    *Sheldon K. Friedlander*

Engineering Our Way Out of Endless Enviornmental Crises .......... 182
    *Walter R. Lynn*

The Paradox of Technological Development ........................ 192
    *Paul E. Gray*

Contributors ..................................................... 205
Index ............................................................ 211

# Technology and Environment: An Overview

JESSE H. AUSUBEL, ROBERT A. FROSCH, AND ROBERT HERMAN

*Be content at least with the verdict of time, which reveals the hidden defects of all things, and, being the father of truth and a judge without passion, is wont to pronounce always, a just sentence of life or death.*

Baldesar Castiglione, The Book of the Courtier, *1528*

What will be the verdict of time on the man-made world? Uneasiness prevails in our newspapers, political forums, and cities; in the forests; and beside the lakes and oceans. Many feel that economic, technological, and scientific developments are accompanied by ever-larger risks for environment, society, and health. With each year, unanticipated and unintended consequences of mature technologies reveal themselves more clearly and long after a commitment to the technologies has suffused the economy: the greenhouse effect from fuels that warm and transport us; the hole in the ozone layer from chemicals that cool our refrigerators and make worries about safe and convenient home food supply a dim memory of grandparents; lung cancer associated with asbestos fibers that were a breakthrough a few decades ago for fireproofing ships, schools, and office buildings. It is equally feared that emerging technologies, such as the genetic engineering of new organisms, will release more problems than they solve. Yet, traditional optimism remains widespread that innovations will be found to finesse or counteract harmful environmental consequences of the ways we transform the planet. Will the verdict on our realization of technology in the environment be life, "sustainable development of the biosphere," or the decay and self-destruction that is one of the futures always envisioned for humankind?

This book seeks to contribute toward answering this question. It articulates what Paul Gray calls the paradox of technology, that environmental disruption is brought about by the industrial economy, but that advancement of the industrial economy has also been and will be a main route to environmental quality. The book examines several analytic frameworks for exploring interactions of technology and environment. It includes review of the history of environment as affected by technology. It offers several technological opportunities to reduce or bypass both current and forecast environmental problems. It provides discussion of social and institutional aspects of the question, for example, how education and the professions must change to play more positive roles in environmental matters. This opening chapter synthesizes the contributions that follow and seeks to place in perspective the relationship between technology and the environment that is the subject of the volume.

Perhaps it is best at this early stage to remind ourselves of some of our technological successes with respect to environment in the broadest sense. It is technology, above all, that has denied or forestalled the original Malthusian vision of population outrunning subsistence. Mankind has been able to modify and increase the size of its niche and sustain increasing population at higher levels of economic well-being. That niches keep changing, through the introduction of new technologies, and that we can change them are too commonly overlooked. For example, systems of transportation and energy have arisen over the past two centuries that would have been unimaginable, given a static definition of resources and unchanging policy with regard to disposal of wastes. The problem of typhoid was largely solved by chlorination of municipal water supplies, although private well owners waited a long time to adopt this solution. In the industrialized world, air and especially water in numerous urban areas are cleaner and safer than a century, or even a couple of decades, ago.

The contemporary question is whether humans may now be so threatening the boundary conditions of the earth system that our technological tool kit will not suffice. Are we infinite or are we reaching closure? We pushed back the North Sea and built more than half the land that is today the Netherlands. Now we wonder whether we dare push nature any further. We drained the malarial marshes of the Maremma on the Italian coast to make them humanly habitable. Now we define global habitability to include many species besides our own.

We must also recognize that many environmental problems have not proven to be as serious as originally forecast. Public alarms about mercury in swordfish, pesticides in cranberry bogs, and radiation from Three Mile Island are among numerous examples. The lesson from these episodes is not that we should distrust all news of environmental dangers, but rather that the public wants a sense of security.

Then what do we learn when we search the history of environment and technology for guidelines? Generally in the industrialized countries, ways have been devised to accommodate and prepare the way for economic growth and increases in population density without decline of key measures of environmental quality and health. Will our ingenuity, technical and social, match current and future needs?

In fact, as Thomas Lee and other authors point out in this volume, both resources and environment are functions of technology. Concerns about scarce resources have repeatedly subsided as technology expanded the available reserve or provided alternatives. According to Lee, the pressure for closure of the system stems more properly from concerns about the capacity of the environment as a receptacle for wastes than from its bounty of resources. It is rarely true that depletion of resources is the driving force for resource substitution. From a historical perspective, energy substitution, for example, has been driven by the availability of a set of new technologies that enabled an alternative energy source to satisfy better and at an acceptable cost the end-use demand of society.

It is useful at this point to distinguish several sources of environmental problems. Some problems come about largely because of irresponsible or unintelligent behavior. Careless ship operations appear to be the immediate cause of the *Valdez* oil spill in Alaska; oil leaks from drilling in the Santa Barbara Channel off California in 1970 could very likely have been prevented by more thorough geological studies and better engineering practice. Some problems arise because of collective effects of individual behavior that is not particularly serious on a small scale or in a forgiving geographical context. The smog of Los Angeles is caused by the sum of a multitude of actions that might be permissible elsewhere, but not in the Los Angeles basin with its enclosing mountain ranges, prevailing westerly winds, and large concentration of people and vehicles. Other problems arise simply out of ignorance. No environmental impact statement at the time of the innovation is likely to have identified the problems that arose decades later with DDT or chlorofluorocarbons (CFCs). Electric refrigeration looked like a marvelous advance over the icebox when it was introduced into the mass market in the late 1920s, and the CFCs looked attractive compared with the problems of leaks and explosions associated with ammonia and other first-generation coolants. Certainly no chemist could have been expected in the 1930s to link CFCs to destruction of stratospheric ozone, which could not be measured accurately at that time, or to the greenhouse effect, then a theory discussed only in the most hypothetical terms by basic scientists interested in the earth's geological history.

In the United States, as Victoria Tschinkel describes, there has been a tendency to treat all kinds of problems the same way, litigiously, and to use a great deal of social effort in attributing effects to causes and assigning blame. It is necessary to recognize better in the U.S. administrative and legal systems that one is not necessarily a horrible individual if one truly did not understand certain things. This volume makes the point strongly that the essence of the environmental crisis is not nearly so much bad actors as the whole, often contradictory, structure of incentives of the economy. Given how complete definition of environmental problems has become (see Table 1), perhaps in the United States for many environmental matters it is time to think more broadly and pragmatically in terms of a "no-fault" society. There is a need to shift from negative to positive reinforcement and to reduce the expense and time involved in resolving disputes. Products and incentives should be designed in such a way that a minimum of hazardous waste is created, but also, it should be easy to dispose of those wastes that are created; society might better use its resources to buy and recycle these materials than to prosecute those who dump them.

A no-fault orientation does not deny the existence of criminality or conflict. On the contrary, we must accept that there are often genuine conflicts of interest on environmental issues, conflicts between industrial and neighborhood objectives or between local and global interests. However, it is becoming rarer for a purely local solution to endure. Globally approaching environmental closure means that, increasingly, we must seek policies that are consistent at all levels of the system and internationally, for example with regard to waste disposal or greenhouse gas emissions.

A no-fault orientation also does not diminish attention to the roles and responsibilities of industry. However, as this volume makes clear, environmental analysis and regulation have sometimes tended to focus on industry as the major force shaping the evolution of the environment to the exclusion of other important forces. And, we have tended to view industry as a collection of pollution sources. As pointed out by Robert Ayres, Sheldon Friedlander, and Robert Herman and coauthors, this view is inadequate. We must be at least as concerned with the environmental consequences of consumption. First we looked at factories, then at some of their products. Now we must encompass the entire system of production and consumption, the metabolism of our society, in our analyses and policies.

## FRAMEWORKS FOR ANALYSIS

In Part 1 of this book, the authors advance three ways of approaching the definition and assessment of environmental problems. The concept of *industrial metabolism* leads to more unified, continuous, and comprehensive consideration of production and consumption processes from an

TABLE 1  Selected Environmental Problems

1. Urban air pollution
2. Regional air pollution, including acid rain
3. Hazardous or toxic air pollutants
4. Indoor radon
5. Indoor air pollutants other than radon
6. Radiation other than radon
7. Depletion of stratospheric ozone associated with CFCs and other substances
8. Global climate change associated with carbon dioxide and other greenhouse gases
9. Water pollution associated with direct and indirect point source discharges from industrial and other facilities to surface waters
10. Water pollution associated with nonpoint source discharges to surface waters
11. Contaminated sludge
12. Pollution of estuaries, coastal waters, and oceans from all sources
13. Deterioration of wetlands from all sources
14. Pollution of drinking water as it arrives at the tap from chemicals, lead in pipes, biological contaminants, and radiation
15. Pollution of groundwater and soil at hazardous waste sites, both sites with continuing disposal and those no longer in use
16. Pollution of groundwater and other media at nonhazardous waste sites, including municipal landfills and industrial sites
17. Exhaustion of landfills
18. Groundwater contamination from septic systems, road salts, injection wells, leaking storage tanks, and other sources
19. Wastes and tailings from mining and other extractive activities
20. Accidental releases of toxic substances
21. Oil spills and other accidental releases of environmentally damaging materials or substances
22. Pesticide residues on foods eaten by humans and wildlife, and risks to applicators of pesticides
23. Risks to air and water from pesticides and other agricultural chemicals as a result of leaching and runoff, aerial spraying, and other sources
24. New toxic chemicals
25. Undesirable environmental release of genetically altered materials
26. Exposure to consumer products
27. Worker exposure to chemicals
28. Reductions in biodiversity
29. Deforestation and desertification

NOTE: These environmental problems can be grouped or ranked according to a variety of criteria, for example, scale (local to global); whether the problems relate primarily to human health or to ecosystems; carcinogenicity; extent to which technical solutions are currently available; and economic costs.

SOURCE: After U.S. Environmental Protection Agency (1988).

environmental point of view. The question of *dematerialization* forces reconsideration of the origins and solutions of environmental issues and places proposals for waste reduction and recycling in context. The examination of *long-term regularities in technological development* provides quantitative evidence of the role of technology in the evolution of environmental problems and offers some optimism about prediction of future problems and their solutions. All of these frameworks might be regarded as elements of a more complete *industrial ecology*, examining the totality or pattern of relations between economic activity and the environment (Frosch and Gallopoulos, 1989).

As described by Ayres, industrial metabolism encompasses both production and consumption, the entire system for the transformation of materials, the energy and value-yielding process essential to economic development. Application of the industrial metabolism viewpoint involves detailed accounting of the flows of materials and energy through human activities. It has yielded a number of important insights.

One insight is that in many places the major human sources of environmental pollutants have been shifting from production to consumption processes. Several industries have increasingly been able to control the materials flows in their production processes quite comprehensively. The history of the chemical industry, for example, is in considerable part one of finding new uses for former waste products. It is probably safe to say, according to Ayres, that industry in the next century will recycle or use a number of today's major tonnage waste products, notably sulfur, fly ash, and lignin waste from paper manufacture.

A second insight is that a large number of materials uses are inherently dissipative. Many materials are degraded, dispersed, and lost (to the economy) in the course of a single normal use. In addition to fuels and food, this applies to many packaging materials, lubricants, solvents, flocculants, antifreezes, detergents, soaps, bleaches and cleaning agents, dyes, paints and pigments, most paper, cosmetics, pharmaceuticals, fertilizers, pesticides, and herbicides. Most of the current consumptive uses of toxic heavy metals, such as arsenic, cadmium, chromium, and mercury, are dissipative in this sense. Other uses are dissipative in practice because of the difficulty of recycling such items as batteries and electronic devices. Increasing product and materials complexity may also contribute to a tendency toward dissipative use, because recycling may become inherently more difficult with complexity.

Thus, although it is important to ask whether in some ways the environmental system is reaching closure, it is also important to recognize that often what must be traced are pathways that are not cycles in a meaningful sense. Although materials do not leave system boundaries, many follow a unique, nonrepetitive evolution on human time scales, combining, recombining, and moving. According to Ayres, more than 90 percent of the total mass of environmentally "active" materials processed annually are converted into waste almost as fast as extracted. It would be useful to develop measures of dissipation and sort out more clearly what can be described accurately as cyclical and what cannot. Finally, it is clear that a significant fraction of materials streams arising from consumptive, dissipative uses is not regularly monitored or perhaps amenable to monitoring. A new vocabulary is needed, emphasizing transformation, transport, and redeposition, and perhaps new indices of dilution and concentration.

From a policy point of view, there are several important consequences of the metabolism perspective. Already noted is the need to attend more to consumption and to develop new concepts for monitoring. Another point is that an effect of dispersion and dissipation of materials is to make problems global. Although problems of production may tend to be industrial and local, problems of consumption will tend to be problems for everyone and global. Ayres also points out that, whereas residuals tend to disappear from the market domain, where everything has a price, they do not disappear from the natural world in which the economic system is embedded. Thus, many signals given by prices are wrong from an environmental viewpoint. For example, differences in prices of coal, oil, and gas scarcely reflect the different environmental consequences of these energy sources.

Industrial metabolism is not a complete model, but it is clearly a useful heuristic device. It makes us more sensitive to comprehensive examination of sources, transport, and fate of pollutants and can lead to earlier identification of problems and a broader range of monitoring, including technological and socioeconomic trends, as well as traditional environmental indicators. It would be desirable to extend the detailed case studies of industrial metabolism beyond those already performed on cadmium and chromium to several other "metabolically active" elements. A need and potential exist for a more systematic look at the material and energy flows of alternative industrial metabolisms, for example, one centered more on use of hydrogen as an energy carrier. The metabolic metaphor is also useful in that it spurs us to think jointly of the health of the ecological and human systems and to look for diseases and treatments. Modifications are clearly needed to increase reliance on regenerative and sustainable processes and to increase efficiency with regard to production and use of by-products.

The term *dematerialization*, explored by Robert Herman, Siamak Ardekani, and Jesse Ausubel, is employed to characterize the decline over time in weight of materials used in industrial end products, or in the "embedded energy" of the products. Dematerialization would be tremendously important for the environment, because less material could translate into smaller quantities of waste generated in both production and consumption. Statements about trends toward dematerialization have been made casually, and these authors seek to provide a systematic basis on which to identify the forces and measures that would allow a credible statement to be made about dematerialization.

There are widely held perceptions of a long-term trend of decline in weight (intensity) of materials and energy embodied in a range of end-use services. Among the evidence pointed to are the decline in per capita consumption of such basic materials as steel in some advanced industrialized countries and the increasing efficiency of energy use. The significant decline in use of steel in the automotive industry does provide strong evidence in support of dematerialization in production. Further evidence of dematerialization in production is provided by data on overall industrial solid waste generation, which showed a significant decline for several years beginning in 1979.

However, the overall picture about dematerialization is not so sanguine. Generation of municipal solid waste has been on the increase, and there appears to have been overall a linear increase in discards with time measured by weight. The potential factors that are offsetting the efficiency gains are numerous. If smaller, lighter products are also inferior in quality, then more units would be produced and the net result could be a greater amount of waste generated. Spatial dispersion of the population is a potential materializer. Migration from urban to suburban areas, often driven by affluence, requires more roads, more single unit dwellings, and more automobiles. The shift from larger families to smaller nuclear families may be a materializer. So may be such activities as photocopying and advertising; the high cost of repair; styles, fashions, and fads; and product innovation. Of course, economic and population growth are major underlying forces.

Herman and coauthors review a number of examples, including the effect of the information revolution on materials demand and waste. Contrary to expectation, the information revolution has led to a significant materialization, especially with respect to paper. In 1959 it was believed that 5,000 Xerox machines would saturate the U.S. market. Instead, in the information era, trees are at risk.

Considerably smaller amounts of waste are generated by most countries with incomes comparable to the United States. The difference is often attributed to more serious effort to recover and reuse wastes, but in fact the differences are not well sorted out. Moreover, the question of dematerialization interacts in complex ways with objectives for system closure or recycling. For example, substituting plastics for steel in a car may reduce weight and increase fuel efficiency but also decrease possibilities for recycling of materials. A question of utmost importance remaining to be addressed is that of rates and styles of materialization of the three-quarters of the world's population in developing countries.

The concept of dematerialization forces evaluation of economic growth in terms that are significant for numerous environmental problems, especially those associated with solid wastes. Like industrial metabolism, dematerialization shows the relative unimportance of production processes per se. The traditional view has been that what leaves the factory gate is good. Dematerialization directs industry and all of society to be more concerned about the eventual fates of its manufactures. It confronts us with the question of whether society can truly afford to continue functioning in its present "throwaway" mode of products such as diapers, batteries, paper, and beverage containers. It suggests that perhaps minimum volume over a product life cycle should be an environmental design criterion, along with factors such as toxicity, and that incentives must be found for cradle-to-grave materials monitoring and responsibility.

Ausubel shows intriguing, though still tenuous, evidence of long-term regularities in the evolution, diffusion, and replacement of several families of technologies that are critical to the environment, including energy, transportation, and materials. Many diffusion processes appear to occur according to a rather strictly set clock. Regular behavior is exhibited over a range of time scales, but what is most impressive is the steady evolution of large systems over periods of many decades. Ausubel does not comment on causes of the behavior but simply points out that the evidence of significant regularities in technological change is increasingly well established.

An example is in pulses of growth in use of energy lasting 40 years or more. There have been at least two of these: one evidently stretching coal to its limits as a fuel and a subsequent pulse in which oil exhausted many of its opportunities in the market. It may be speculated that during each pulse the leading source of energy supply reaches environmental and other constraints that limit the overall growth of the energy system. In other words, a characteristic density of use may be all that is achievable or socially tolerable for each source of energy within the context of the larger industrial paradigm in which that source of energy dominates. To accommodate further increases in per capita energy consumption, each time it is necessary for a society to shift to a source of primary energy that is not only economically sound but also more environmentally compatible and in some ways more efficient, especially in transport and storage.

More generally, the focus on long-term evolution of technology highlights many remarkably positive aspects of the performance of engineering with respect to resources and environment over the past century. Series of innovations have been brought forth to escape what appeared to be insoluble problems of shortages of resources (such as wood for railroad ties) or overburdening of the assimilative capacity of the environment (for example, waste from the growth of the population of horses in cities around the turn of the century).

The existence of long-term regularities may have predictive value if we observe them early enough and can estimate their characteristics. Ausubel's chapter suggests that an era may be under way in which it is possible to predict with greater accuracy and reliability the emergence of environmental problems. It might provide new perspectives for setting priorities among environmental problems. The questions would be, What technologies are most promising in light of what is understood about overall trajectories of technological development? Can policies be implemented that will enhance the diffusion of selected technologies? How quickly and to what extent can technologies be deflected from well-established trajectories?

Together, the three related frameworks for analysis described above promise to provide a much stronger foundation for our understanding of the technological sources of environmental change. Such a foundation is essential for development of projections of future loading of the environment in which we have more confidence. These projections increasingly form the basis of both social regulation and environmental research.

## SOME SOLUTIONS

In Part 2, Richard Balzhiser and Thomas Lee propose some technological contributions to solutions of environmental problems associated with energy production and consumption. There is general agreement that reduction in emissions from the supply side and improvement in efficiency on the demand side are the right things to do. For the supply side, the technological tool kit appears to be well stocked, for example, to burn coal much more cleanly to alleviate problems of acid rain. Indeed, the record of engineering achievement shows sustained improvement in thermal efficiency accompanied by a continuing decline in the cost of electricity over most of the century. In the past few years, energy requirements and losses associated with stringent emission controls have offset continued engineering improvements aimed at efficiency.

An immediate task is to find the next generation of technology that exploits a basically different systems approach to clean coal combustion. However, this may be only a local or short-run solution, because it may increase the carbon dioxide emissions associated with global climate warming. From this point of view, natural gas is the most convenient fuel of choice for addition and replacement of electricity-generating capacity for the next decade.

Gas is clean and available, and it minimizes exposure to financial risk in investment. It is convenient and also quick to install in relatively small increments for either utility or nonutility generation and for cogeneration. At current prices and with available technology gas is the option of economic choice in the United States not only for peaking but also for middle-range and some base-load applications. It offers the prospect of continuing technical advances in the lifetime and efficiency of gas turbines and in combined gas-steam cycle systems.

From the perspective of long-term regularities in the energy system, gas also appears to be the fuel of choice. It is on a vigorous trajectory toward increased market share. Moreover, gas technology is still young. For years, natural gas was a by-product of oil exploration. Only recently have many wells been drilled intentionally for gas exploration. Effectiveness in gas exploration is growing by use of satellite remote sensing and ground truth measurements, and drilling technologies are advancing underground to greater depth with increased speed and accuracy.

To optimize further use of coal as well as gas resources, the integrated gasifier with combined cycle (IGCC) can be considered a major step forward. If carbon dioxide must eventually be removed from power plant effluents, IGCC can probably best accommodate this requirement, not without cost, but at costs below other coal-based alternatives. Meanwhile, gas produces less carbon dioxide per kilowatt-hour than any other fossil fuel option and permits us some time to understand better the issue of climate change without imposing costly but ineffectual carbon dioxide removal requirements.

At present, the United States seems to be adding boundary conditions to the energy industry in such a way that Balzhiser predicts market shares of primary energy sources for generating electricity will remain almost perfectly static as far as the year 2020. This seems a most unlikely development, given the patterns of change over the past 100 years. Moreover, no one believes that the United States is now at an economic or environmental optimum in the energy sector. However, Balzhiser points out that most energy decisions in the United States are being postponed, only small incremental changes are being made, and thus the key choices are arising by default. As with the rest of the aging national infrastructure, the United States is taking the energy infrastructure for granted and living off investments of the past. In this circumstance, extension of plant lifetimes has become one of the most important engineering challenges. This involves both sensor technologies and computer aids that give much broader coverage of equipment with on-line diagnostics.

Lee and Balzhiser envision an evolution of the energy system to one that integrates energy into a system of materials processing, a realization of the "metabolic" view proposed by Ayres. It might begin as a marriage of coal and gas technologies and evolve ultimately into fully "integrated energy systems" (IES). The IES concept is one in which product streams and energy streams merge. The increasing orientation of power plants toward chemical process is taken seriously in all dimensions. Coal, crude oil, liquefied petroleum gases, and natural gas could all be primary materials used by the system. For example, natural gas would be used as a fuel in heaters, as a feedstock, or as a fuel for making hydrogen. Intermediate industrial gases are exploited to their maximum benefit. The entire steam system of the facility is integrated and, in turn, integrated with the electric system. Waste of heat or components is minimized, thereby enhancing economic efficiency. Zero emission, the ultimate dream for energy systems, as Lee points out, can be accomplished only with a hydrogen economy, and IES offers a technological road toward that goal.

Depletion of the ozone layer is another illustrative tale of technology and environment, as described from the perspective of U.S. industry by Joseph Glas. Chlorofluorocarbons were invented around 1930 as a safe alternative to ammonia and sulfur dioxide for use in home refrigerators. The intent was to eliminate the toxicity, flammability, and corrosion concerns of the other chemicals by developing a stable chemical with the right thermodynamic properties. That effort was so successful that the new compounds were also quite easy to make and rather inexpensive. New applications for a safe class of chemicals with the properties of CFCs were plentiful, and the market blossomed. Currently, virtually all refrigeration, commercial air-conditioning, defense and communications electronics, many medical devices, and high-efficiency insulation use CFCs in some way. But today, more than 50 years after the development of CFCs, we have modified and extended our definition of "safe."

It now seems likely that CFCs will be largely phased out over the next 10–20 years, and it is the development of new technologies that has provided what appear to be viable options for meeting society's demands simultaneously for caution on environmental modification and for the services provided by CFCs. In the extreme, a ban on CFCs before alternative chemicals or technologies can be put into place would be damaging both to safety and to economics. Moreover, it would almost certainly be ineffective. Unified action and implementation on a global scale are needed, and bans in the absence of alternatives would likely lead to uncontrollable, uncooperative behavior both by producers still seeing a market opportunity and by consumers wanting a service.

In the ozone story, the rate of technological progress and the degree of risk are inextricably related. It is a story that promises a gratifying outcome, with science, governments, and industry acting forcefully by building on common goals of protecting the environment and agreeing on technical analyses.

## SOCIAL AND INSTITUTIONAL ASPECTS

Why is the promising story of stratospheric ozone protection not repeated more often? A key reason, articulated by Tschinkel in Part 3, has been the inadequacy of individual professions in the face of complex problems and an equal difficulty among key groups in making common cause. As Tschinkel describes, a succession of professions in the United States has discovered environmental problems over the past 100 years: first physicians and experts in public health, then engineers, later biologists and toxicologists, and most recently lawyers. Each discovered problems and offered solutions. All of the solutions have had unforeseen consequences, whether natural, social, or economic, and most have been so narrowly focused that opportunities to achieve larger public goals have been missed or obscured.

Given the elusiveness of assigning causes, predicting effects, and finding cures for many environmental problems, it is not surprising, as Tschinkel points out, that the United States has developed a condition ripe for the legal profession to flourish in. In the past 20 years the legal system has generated some of the key decisions supporting environmental protection. As Tschinkel contends, it has also produced an adversarial, combative climate in which it is virtually impossible for people from industry to discuss facts with their colleagues in government or the public. We "are constantly in litigation and constrained from solving problems by using each other's talents cooperatively." Moreover, the litigation is often not fruitful. For those environmental cases that went to trial in federal civil courts, 10 percent took longer than 67 months to resolve. Most serious, according to Tschinkel, the legalistic approach has produced a staggering load of regulations that leaves little time or incentive for creativity and human judgment in developing solutions and no time for concentrating on environmental results. It has created a process-oriented, rather than a results-oriented, approach in a sector where the result, namely, environmental quality, is what we seek and need.

In fact, the succession of legislative activities has resulted in an enormous, sometimes contradictory, uncoordinated patchwork of control requirements for smoke, air and water pollution, solid wastes, and noise, as well as aesthetics. An example is that U.S. regulations require advanced waste treatment of domestic waste at about 50 percent higher cost than the usual secondary treatment when discharged into a eutrophic water body. Next to this "gold-plated pipe" is often found a storm water ditch carrying the equivalent of raw sewage: the water that flows through it receives absolutely no treatment. What strategy makes sense in this situation for technologists and, indeed, for society as a whole? On the one hand, better engineering would create fewer problems for biologists and lawyers to worry about; on the other hand, imaginative approaches are needed to foster cooperative activity between technical experts and the policymaking community.

As Friedlander stresses, the technological community, indeed all of society, has been largely reactive to environmental issues. In the past we have tended to wait for crises, as Walter Lynn chronicles, and then responded. Society needs a positive agenda for environment, based on more comprehensive theories, better data bases, and better analyses. For engineers, the emphasis should be on design of environmentally compatible technologies, both for manufacturing and plant operations and for products. The latter must be accented, while evidence grows that environmental consequences of consumer products may be more important than the direct effects of industrial activity, as demonstrated by the perspectives of industrial metabolism and dematerialization.

Design should not merely meet environmental regulations; environmental elegance should be part of the culture of engineering education and practice. Selection and design of manufacturing processes and products should incorporate environmental constraints and objectives at the outset, along with thermodynamic and economic factors. Ever-increasing goals for environmental quality present the engineering profession with challenges in design, basic research, and education.

Environmental engineering must become a more integral part of chemical, manufacturing, materials, and other engineering fields, not only civil engineering, to which it is traditionally closest. Environmental quality must become a pervasive ethic in all engineering design. In turn, values must be transformed into engineering requirements—values about preservation of ecosystems and biota, protection of public health, and intergenerational responsibility. In Gray's words, "the great hope and the great challenge before us are to bring engineering education and practice, industrial priorities, and public policy into alignment in ways that eliminate the paradox of technological development."

Over the past few years, as described by Friedlander, a movement has grown stressing design and in-plant processes, in contrast to add-on devices or exterior recycling, to reduce or eliminate waste. This movement has been called waste reduction or pollution prevention. Current regulatory practices focus almost exclusively on what comes out of the pipe or smokestack. They ignore broader systems-oriented approaches and the assimilative capacity of the environment, and impose lockstep application of selected technologies.

End-of-the-pipe approaches will provide few further benefits. We need first to prevent waste creation. This involves the development of substitute products and processes emphasized by Ausubel and reengineering much of what is done in key industries. We should seek general principles to guide the search for substitutes for certain broad classes of widely used materials with environmental effects. As mentioned by Friedlander, in response to developing regulatory trends and competition from the paper industry, the chemical industry in the United States and Europe has begun development of biodegradable plastics, much as was done 25 years ago, when long-lasting detergents were polluting water bodies.

It is important to gain acceptance of the primacy of reducing waste and preventing waste creation. The primacy rests on several factors, identified by Friedlander. Avoiding the creation of a waste eliminates the need for its treatment and disposal, both of which carry environmental risk. Control technologies may fail or fluctuate in efficiency. Treated effluent streams carry nonregulated residual substances that may turn out later to be harmful. Secured disposal sites eventually discharge into the environment.

Methods of waste reduction include in-plant recycling, changes in process technology, changes in plant operation, substitution of input materials, and modification of end products both to permit use of less-polluting upstream processes and to prevent the products themselves from becoming problem wastes. According to Friedlander, the technology of waste reduction does not yet have a widely accepted scientific basis. There is a need to find a class of generic scientific and engineering principles that will eventually make it possible, in the words of Tschinkel, for the concept of treatment to become passé.

In the meanwhile, it is desirable to follow a clear hierarchy in waste management (Science Advisory Board, 1988). If waste cannot be prevented, then we should seek to recycle or reuse it. However, recycling may have acquired a level of visibility as a potential solution that exceeds its promise. Apart from behavioral and economic hurdles, recycling faces technical limitations. For example, recycling paper shortens paper fibers and lowers quality. There are precious metals, such as platinum and rhodium used in catalytic converters, that industry would like to recycle, but an economic means to collect the converters has not yet been found. It must, moreover, be recognized that many recycling sites have subsequently become "Superfund" sites, where cleanup activities are required.

If recycling or reuse is not possible, then it is time for treatment and destruction, relying on technologies such as bioremediation and incineration. If those options are insufficient, the next resort is waste isolation, for example, well-constructed sanitary landfills. The last resort is avoiding exposure to released residues. It is important to point out that even with waste reduction, incineration, and recycling, no landfills will remain in a couple of decades or sooner for many major population concentrations. Globally, and especially in the industrialized countries, we are faced with our own materialization, a culture that in the United States produces some 5 to 10 pounds of waste per capita per day, depending on the comprehensiveness of definition of the term. Even with remarkable engineering achievements, many of the problems associated with waste disposal will become worse.

Although it may not be possible to represent fully costs relating to health and ecosystems in economic terms, a key need is the economic data base to support decision making about waste reduction and alternatives.

Even in a prominent case like automotive tires, there is no strong analytic base for evaluating the relative merits of gradual decomposition versus burning or smelting. More effort is required to calculate the true costs of waste disposal options, including potential liability costs.

## TECHNOLOGICAL OPPORTUNITIES

In general there is a need to identify, research, and put into practice high-leverage areas of innovation for environmental quality. Already mentioned is the need for biodegradable plastics; more could be understood about using ultraviolet radiation or gamma rays to irradiate and harmlessly decompose plastics. Materials research itself can be a key to dematerialization. More needs to be understood about incineration and combustion; progress on fundamentals of combustion is already enabling the design of engines that produce lower $NO_x$ emissions. Microbial transformation of wastes, for example, selective removal of heavy metals, offers promise. It is time to become serious about technologies for reducing and recycling carbon dioxide emissions. Technologies for cost-effective separation of hydrogen remain areas of potentially high environmental payoff. There are also pervasive needs for improvement and deployment of monitoring technologies. Environmental monitoring remains labor intensive and based on technologies that should soon be superseded by new sensors and measurement methods.

As Gray argues, the growing concentrations of greenhouse gases in the atmosphere logically lead to a reconsideration of the possibility of increasing the use of nuclear energy. Gray proposes that we develop, build, and test radically different reactor designs that pose negligible risks of the accidental release of radioactive materials as a result of overheating. Several possibilities exist, including new water-cooled and liquid-metal-cooled designs, as well as gas-cooled designs. These hold the promise of passively safe operation. The nuclear question is a reminder that many engineering systems have been poorly designed from the point of view of operators and that this human aspect of design must be taken more seriously, whether in electrical or chemical plants, supertankers, or consumer products.

One of the most interesting questions is that of research and market opportunities with regard to efficiency, especially energy efficiency. In the past few years there has been a shift among many environmentalists to a revised view of the "soft path" option that emphasizes managing demand downward rather than supply upward to meet societal needs and problems. The revised view emphasizes efficiency but omits life-style changes that were part of the soft path program in the 1970s.

Still, why is efficiency gaining much less than predicted and espoused? There may be several answers. One is almost certainly the answer usually proposed by soft path advocates, namely, that the playing field is not level for the competition between conservation technologies and supply-expanding technologies because of tax and other subsidies available to such industries as oil exploration. Another may be that energy, and economic growth as traditionally defined, remain a solution to many problems. As an example, Ausubel points out that catalytic converters reduce the energy efficiency of cars, although they serve other highly desirable environmental objectives.

At a deeper level the problem may be that end-use efficiency is almost always the result of a process involving several links in a chain. Ayres points out that a new process that saves one link in the chain between raw materials and final goods or services can usually be justified in terms of savings in raw materials, energy, or capital requirements. Final products are made by sequences of processes with an overall conversion efficiency that is the product of the efficiency at each stage. If a typical chain has four steps, each with a very favorable conversion efficiency of 0.7, the overall conversion ratio of the chain (i.e., $0.7^4$) is about 0.24. The world energy system appears now to have an overall efficiency of only about 0.15. We have been climbing the overall efficiency curve steadily but slowly, on average about 2 percent per year over the past 100 years, a rate requiring 35 years for a doubling of performance (Grübler and Nakicenovic, 1989). The promise of energy efficiency is there, but the basic structural problem may yield only gradually. Meanwhile, it makes sense to seek gains at each link in the chain and to pursue efficiency technologies that would both moderate demand and deliver supply more cost-effectively. The latter would include research into such areas as superconducting cables for electricity transmission, magnetic induction for motors and transportation, and microwave heating.

As Friedlander points out, there should be a large, profitable market for technologies that are both environmentally and economically competitive. Inherently safer and less-polluting plants should cost less for several reasons. Presumably, prevention of pollution should pay for itself through reduced demand for inputs and reduction in waste disposal and liability costs. Moreover, if products can be made smaller and lighter without loss of quality, there may be an economically attractive reduction in space required for manufacturing and storage and in transportation costs.

A critical question is on what terms wealthier countries will transfer low-pollution technologies to developing countries. Friedlander points out several consequences of worldwide increases in pollutant emissions that will occur as more nations industrialize:

- pressure for further regulation at the international level,

- difficult international negotiations aimed at setting national emission allocations, and
- competitive advantage to nations whose engineers are able to design clean, economic technologies.

We may be entering a new era of complex global bargaining, where environmental quality is a major objective and where environmentally attractive technologies and resources are major bargaining chips. What terms will be required to prevent the cutting down of tropical forests or the building of new coal-burning power plants? Environmental protection on a global scale will require the nations with low-pollution technologies to transfer these in a timely fashion in both subsidized and unsubsidized ways.

The research agenda must recognize that we face complex, interdisciplinary problems, that systems research is required, that we should acquire as much information as possible at a minimum within reasonable cost boundaries, that we should address these with operational and social science as well as engineering and natural sciences, and that many points of view should be folded in. We should strive to go from particular problems toward general analytic understanding and global methods. The lack of training in methods for solving complex problems is evident in the history of environmental policy, as Tschinkel illustrates.

Yet, the partial nature of solutions must be accepted. On most environmental issues, the luxury of time to search for optimality does not exist, as Glas's account of the ozone controversy demonstrates. In many areas, decisions are being made in real time or on the basis of anticipated consequences. However, a sequence of partial solutions may form a good path if the forces driving the system are reasonably well understood. The key is to work on narrow or specific problems with an understanding of the interface with the overall problem; this is well illustrated by questions surrounding clean coal technologies (which could alleviate acid rain) and CFC substitutes (which might not destroy ozone) but could intensify concerns about greenhouse warming. Isolating meaningful subsystems is not always easy, especially in turbulent, dynamic systems such as those in which most environmental issues exist. Still, it is necessary to take the long and dynamic view and to build the research capital and data base. Moreover, as Ausubel points out, there may occasionally be simple and important relations to be found in complex systems. The challenge is to couple technological, economic, and environmental considerations without unrealistic data needs and analytic paralysis that can come with overly ambitious modeling efforts.

## CONCLUDING THOUGHTS

Some of the toughest issues facing us—urban air pollution, climatic change, destruction of rain forests, and loss of habitat, for example—are the consequences of large-scale cultural patterns, the summed effects of millions of people making individual decisions. Engineering communities have become painfully aware that such phrases as the "tragedy of the commons" and the "tyranny of small decisions," described by Lynn, are accurate descriptions of reality, a reality that is most difficult to alter. What will be needed to make more people work together and act for the common good? Gray suggests that we are caught in a gridlock of adversarial relations on environmental matters, but only when virtually everyone wants to escape can such social traps be broken.

Certainly there is a need to educate individuals about how behavior in exercising consumer preferences affects local and global environments. In the environmental problem in the large, we confront the desire for a generalized freedom; as individuals, we do not want constraints imposed that affect mobility, life-style, convenience, or purchasing decisions. To quote Lynn, "It is true that regulations reduce our freedom of choice, but so does a deteriorating environment." Tschinkel's formula for progress is to have a sound scientific base and incentives for doing the right thing and to engage people for cooperation early in the process, to recognize that human beings are mere mortals, to be relatively site specific and results oriented, and to seek agreements that are negotiated and approved by all parties.

As Ayres points out, contrary to popular belief, long-term goal orientation in economic life is not particularly rare. But there is a need to increase shared recognition of a long-run evolutionary imperative that favors an industrial metabolism that results in reduced extraction of virgin materials, reduced loss of waste materials, and increased recycling of useful materials. Although the overall trend may not yet exist, the imperative is to seek reduced materials intensiveness or dematerialization.

Technology should become a ground on which takes place the problem-oriented reintegration of the domains of human knowledge and social development. As Tschinkel reminds us, customary scientific analysis does not show what is detrimental for the environment and what is not; it states what consequences logically follow from what activities. Accelerated by new research programs, natural science will very likely reveal more in the 1990s about nature than mankind has ever known, but scientific analysis as traditionally practiced will remain unseeing with respect to human needs. Meanwhile, the social sciences, especially economics, which in theory address human needs, have proved almost blind with regard to nature. The intersection of technology and environment in a sense has been the blind spot in our system of knowledge, and this gap is at the root of today's environmental crisis (Meyer-Abich, 1979).

Environmental engineering, recognizing our own nature as part of nature and our technology as in nature, can help bridge the dangerous compartmentalization of knowledge and professions that appears to be placing modern life in jeopardy. The technological potential is there for both economic growth and improvement in environmental quality. However, technological and scientific development is embedded from the beginning in social developments, and there we must harmonize incentives to foster a world of environmental and economic quality rather than desolation and self-burial.

We must also accept that in responding to many environmental questions, we may never know whether we are right or wrong even in the case of more narrowly defined scientific aspects of a given issue. Much of the environmental investment that must be made in coming decades will be like the building of the great Gothic cathedrals, performed out of respect for large and durable forces and to address noneconomic concerns. We cannot be certain what would occur if we fail to take precautions. Many of the challenges to be faced, like global warming, exist to a considerable extent in the domain of "hypotheticality" (Häfele, 1975).

So then what will be the verdict on the earth transformed by human activity? We should not simply stand by to find out, but try always to create the technological conditions in which it will be meaningful and satisfying to ask many other questions about human existence.

## REFERENCES

Frosch, R., and N. Gallopoulos. 1989. Strategies for Manufacturing. Scientific American 261(3):144–152.

Grübler, A., and N. Nakicenovic. 1989. Technological Progress, Structural Change, and Efficient Energy Use: Trends Worldwide and in Austria. Laxenburg, Austria: International Institute for Applied Systems Analysis.

Häfele, W. 1975. Hypotheticality and the new challenges: The pathfinder role of nuclear energy. Anticipation 20:15–23.

Meyer-Abich, K. M. 1979. Toward a practical philosophy of nature. Environmental Ethics 1(4):293–308.

Science Advisory Board, U.S. Environmental Protection Agency. 1988. Future Risk: Research Strategies for the 1990s. SAB-EC-88-040. Washington, D.C..

U.S. Environmental Protection Agency. 1988. Unfinished Business: A Comparative Assessment of Environmental Problems. Springfield, Va.: National Technical Information Service.

# 1
# Frameworks for Analysis

# Industrial Metabolism

ROBERT U. AYRES

We may think of both the biosphere and the industrial economy as systems for the transformation of materials. The biosphere as it now exists is very nearly a perfect system for recycling materials. This was not the case when life on earth began. The industrial system of today resembles the earliest stage of biological evolution, when the most primitive living organisms obtained their energy from a stock of organic molecules accumulated during prebiotic times. It is increasingly urgent for us to learn from the biosphere and modify our industrial metabolism, the energy- and value-yielding process essential to economic development. Modifications are needed both to increase reliance on regenerative (or sustainable) processes and to increase efficiency both in production and in the use of by-products.

In this chapter, mass flows for key industrial materials of environmental significance, and the waste emissions associated with them, are reviewed along with the environmental impact of the waste residuals; economic and technical forces driving the evolution of industrial processes; long-range tendencies in the development of the industrial metabolism; and some applications of "materials-balance" principles to provide more reliable estimates of outputs of waste residuals. The Appendix contains theoretical explorations of the biosphere and the industrial economy as materials-transformation systems and lessons that might be learned from their comparison.

Before presenting the main analytic framework, it is useful to begin with some positive examples of how industrial metabolism can shift in the direction of increased efficiency in materials flows and waste streams. It has been justly remarked that the history of the chemical industry is one of finding new uses for what were formerly waste products (Multhauf, 1967). One of the most interesting early examples of such an innovation was the Leblanc process (the predecessor of the Solvay process) for manufacturing sodium carbonate circa 1800. As a raw material, it made use of sodium sulfate, an unwanted by-product of the eighteenth-century process for manufacturing ammonium chloride (sal ammoniac). Sal ammoniac was used for cleaning metals and as a convenient source of ammonia, but sodium sulfate had no use. The Leblanc process reduced sodium sulfate to sodium sulfide by heating it in a furnace with charcoal. This, in turn, was heated with calcium carbonate (chalk), which induced a reaction producing the desirable sodium carbonate and a new waste product, calcium sulfide (Multhauf, 1967).

Meanwhile, the market for ammonium chloride failed, so sodium sulfate had to be produced in the same complex by reacting sulfuric acid with common salt. This yielded hydrogen chloride (hydrochloric acid) as another by-product. In this case, the hydrochloric acid quickly found a practical use in the manufacture of chloride of lime as a bleach for the rapidly growing textile industry. The other waste product, calcium sulfide, was not successfully used until the 1880s, as a source of sulfur for the manufacture of sulfuric acid.

One of the most important waste products of the early nineteenth century was coal tar, which was generated in large amounts by gasworks making "town gas" for illumination. It was a systematic search for useful by-products, initiated by German chemists, that resulted in the creation of the modern organic chemical industry. The synthetic aniline dyes introduced after 1860 were all essentially derived from chemicals obtained from coal tar.[1] Phenolic resins, aspirin, and the sulfa drugs are later examples of derivatives.

Until recently, even natural gas was regarded as a by-product of petroleum production. Although it was used almost from the beginning as a fuel for refinery processes and illumination in local areas, much of it was wasted by "flaring" (indeed, this is still true in remote areas of the world). It had no significant chemical uses until World War II, when natural gas became the feedstock for producing ethylene, and thence butadiene and styrene, the major ingredients of synthetic rubber.

Chlorine is a final example. It is manufactured jointly with sodium or sodium hydroxide (lye) by electrolysis of salt or brine. When this process was introduced in the 1890s, it was the sodium hydroxide that was wanted for a variety of purposes, including petroleum refining and soap manufacturing. At the time, chlorine was a low-value by-product, which was luckily available for use in municipal water treatment. The development of a number of valuable chlorine-based organic chemicals (e.g., the most

TABLE 1 Mass of Active Materials Extracted Commercially in the United States, 1960-1975 (million tons)

| Material | Year | | | |
|---|---|---|---|---|
| | 1960 | 1965 | 1970 | 1975 |
| Food and feed crops (excluding hay) | 267 | 295 | 314 | 403 |
| Meat, fish, and dairy products | 82 | 85 | 85 | 84 |
| Cotton, wool, hides, and tobacco | 5 | 5 | 4 | 4 |
| Timber (15% moisture basis) | 256 | 267 | 271 | 249 |
| Fuels (coal, lignite, oil, gas) | 990 | 1,176 | 1,458 | 1,392 |
| Ores (Fe, Cu, Pb, Zn) | 400 | 435 | 528 | 460 |
| Nonmetallics | 200[a] | 240[a] | 266 | 255 |
| Total | 2,200[a] | 2,500[a] | 2,926 | 2,847 |

[a] Estimated value.

NOTE: Vegetable material harvested directly by animals has been omitted for lack of data, along with some obviously minor agricultural and horticultural products. Figures for metal ores exclude mine tailings and gangue removed to uncover ore bodies. Inert construction materials such as stone, sand, and gravel have also been omitted. Inert materials account for enormous tonnages, but undergo no chemical or physical change except to the extent that they are incorporated in concrete or paved surfaces. The table also excludes soil and subsoil shifted during construction projects or lost by erosion.

SOURCE: U.S. Bureau of the Census (1960-1975).

common industrial solvents and refrigerants, pesticides, herbicides, and plastics such as polyvinyl chloride) actually reversed the position. By the 1950s and 1960s, chlorine was the primary product and sodium hydroxide was the less valuable by-product.

## MASS FLOWS AND WASTE EMISSIONS

Our economic system depends on the extraction of large quantities of matter from the environment. Extraction is followed by processing and conversion into various forms, culminating in products for "consumption." The U.S. economy extracts more than 10 tons of "active" mass per person (excluding atmospheric oxygen and fresh water) from U.S. territory each year. Of the active mass processed each year, roughly 75 percent is mineral and "nonrenewable," and 25 percent is, in principle, from renewable (i.e., biological) sources as shown in Table 1. Of the latter, much is ultimately discarded as waste, although much of it could (in principle) be used for energy recovery.

It is difficult to estimate the fraction of the total mass of processed active materials that is annually embodied in long-lived products and capital goods (durables). None of the food or fuel is physically embodied in durable goods. Most timber is burned as fuel or made into pulp and paper products. At least 80 percent of the mass of "ores" consists of unwanted impurities (more than 99 percent in the case of copper). Of the final products made from metals, a large fraction is converted into "consumption goods," such as bottles, cans, and chemical products, and "throwaway" products such as batteries and light bulbs. Only in the case of nonmetallic minerals (if inert materials are ignored, as before) is as much as 50 percent of the mass embodied in durable goods (mainly portland cement used for concrete and clays used for bricks and ceramics). The annual accumulation of active materials embodied in durables, after some allowance for discard and demolition, is probably not more than 150 million tons, or 6 percent of the total. The other 94 percent is converted into waste residuals as fast as it is extracted.

Combustion and carbothermic reduction processes are the major sources of atmospheric pollutants today but by no means the only important ones; nor is the atmosphere the only vulnerable part of the environment. From a broader environmental perspective, the production and dispersal of thousands of synthetic chemicals—many new to nature, and some highly toxic, carcinogenic, or mutagenic—and the mobilization of large tonnages of toxic heavy metals are of equally great concern. The complexity of the problem is too great to permit any kind of short summary.

However, two points are worthy of emphasis. First, as noted above, most materials "pass through" the economic system rather quickly. That is to say, the transformation from raw material to waste residual takes only a few months to a few years in most cases. Long-lived structures are very much the exception, and the more biologically "potent" materials are least likely to be embodied in a long-lived form.

The second point is that many uses of materials are inherently dissipative (Ayres, 1978). That is, the materials are degraded, dispersed, and lost in the course of a single normal use. In addition to food and fuels (and additives such as preservatives), other materials that fall into this category include packaging materials, lubricants, solvents, flocculants, antifreezes, detergents, soaps, bleaches and cleaning agents, dyes, paints and pigments, most paper, cosmetics, pharmaceuticals, fertilizers, pesticides, herbicides, and germicides. Many of the current consumptive uses of toxic heavy metals such as arsenic, cadmium, chromium, copper, lead, mercury, silver, and zinc are dissipative in the above strict sense. Other uses are dissipative in practice because of the difficulty of recycling such items as batteries and electronic devices.

Tables 2 and 3, which summarize estimates of emission coefficients for various heavy metals by use category and annual emissions attributable to dissipative uses, make clear the heterogeneity of the materials flows. In some cases the dissipation is slow and almost invisible. For instance, paints (often containing lead, zinc, or chromium) gradually crack, "weather," and turn to powder, which is washed or blown away. Tires, which contain zinc (and cadmium) salts, are gradually worn away during use, leaving a residue on the roads and highways. Similarly, shoe leather, containing up to 2 percent chromium salts, is gradually worn away to powder during use. Incinerator ash contains fairly high concentrations of heavy metals from a variety of miscellaneous sources, ranging from used motor oil to plastics and pigments.

On reflection, many dissipative uses (food and fuel again excepted) are generally seen to be nonessential in the sense that technologies are theoretically available, or imaginable, that could eliminate the need for them. To take one example only: hydroponic agriculture in enclosed, atmospherically controlled greenhouses, with genetically engineered pest controls, would eliminate all losses of fertilizers and pesticides to watercourses by way of surface runoff.

## ENVIRONMENTAL IMPACT OF WASTE RESIDUALS

Materials do not disappear after they are used up in the economic sense. They become waste residuals that can cause harm and must be disposed of. In fact, it is not difficult to show that the tonnages of waste residuals are actually greater than the tonnages of crops, timber, fuels, and minerals recorded by economic statistics. Although usually unpriced and unmeasured, both air and water are major inputs to industrial processes and they contribute mass to the residuals—especially combustion products. Residuals tend to disappear from the market domain, where everything has a price, but not from the real world in which the economic system is embedded.

Many services provided by the environment derive ultimately from "common property," including the air, the oceans, the genetic pool of the biosphere, and the sun itself. Distortions in the market (i.e., prices) are unavoidably associated with the use of common property resources. Clearly, environmental resources such as air and water have been unpriced or (at best) significantly underpriced in the past. For this reason, such resources have generally been overused.[2]

As noted, the total mass of waste residuals produced each year by industrial metabolism far exceeds the mass of active inputs derived from economic activities. This is because nearly half of the inputs other than air and water are fossil fuels (hydrocarbons), which combine with atmospheric oxygen and form carbon dioxide and water vapor. The carbon fraction of hydrocarbons ranges from 75 percent in methane to about 90 percent

TABLE 2 Consumption-Related Emissions Factors: Heavy Metals

| Metal | Metallic Use[a] | Painting and Coating[b] | Paint and Pigments[c] | Electron Tubes and Batteries[d] | Other Electrical Equipment[e] |
|---|---|---|---|---|---|
| Silver   | 0.001 | 0.02 | 0.5 | 0.01 | 0.01 |
| Arsenic  | 0.001 | 0    | 0.5 | 0.01 | N.A. |
| Cadmium  | 0.001 | 0.15 | 0.5 | 0.02 | N.A. |
| Chromium | 0.001 | 0.02 | 0.5 | N.A. | N.A. |
| Copper   | 0.005 | 0    | 1   | N.A. | 0.10 |
| Mercury  | 0.050 | 0.05 | 0.8 | 0.20 | N.A. |
| Lead     | 0.005 | 0    | 0.5 | 0.01 | N.A. |
| Zinc     | 0.001 | 0.02 | 0.5 | 0.01 | N.A. |

N.A. = Not applicable.

[a] Alloys or amalgams (in the case of mercury) not used in plating, electrical equipment, catalysts, or dental work. Losses can be assumed to be due primarily to corrosion, except for mercury, which volatilizes.

[b] Protective surfaces deposited by dip coating (e.g., galvanizing), electroplating vacuum deposition, or chemical bath (e.g., chromic acid). The processes in question generally resulted in significant waterborne wastes until the 1970s. Cadmium-plating processes were particularly inefficient until recently. Losses in use are mainly due to wear and abrasion (e.g., silverplate) or to flaking (decorative chrome trim). In the case of mercury-tin "silver" for mirrors, the loss is largely due to volatilization.

[c] Paints and pigments are lost primarily by weathering (e.g., for metal-protecting paints), wear, or disposal of painted dyes or pigmented objects, such as magazines. Copper- and mercury-based paints volatilize slowly over time. A factor of 0.5 is assumed arbitrarily for all other paints and pigments.

[d] Includes all metals and chemicals (e.g., phosphorus) in tubes and primary and secondary batteries, but excludes copper wire. Losses in manufacturing may be significant. Mercury in mercury vapor lamps can escape to the air when tubes are broken. In all other cases it is assumed that discarded equipment goes mainly to landfills. Minor amounts are volatilized in fires or incinerators or lost by corrosion. Lead-acid batteries are recycled.

[e] Includes solders, contacts, semiconductors, and other special materials (but not copper wire) used in electrical equipment control devices, instruments, etc. Losses to the environment are primarily through discard of obsolete equipment to landfills. Mercury used in instruments may be lost through breakage and volatilization or spillage.

in anthracite. Petroleum is intermediate. Every ton of fuel carbon is converted into 3.67 tons of carbon dioxide emitted to the atmosphere.[3] It is thought that only about half of this amount remains in the atmosphere, but the carbon dioxide level of the earth's atmosphere has risen over the past century from about 290 parts per million (ppm) to around 344 ppm at present. Although the magnitude of the climatic warming (the "greenhouse

| Metal | Chemical Uses Not Embodied[f] | Embodied[g] | Agri-cultural Uses[h] | Nonagri-cultural Uses[i] | Medical, Dental[j] | Miscella-neous[k] |
|---|---|---|---|---|---|---|
| Silver   | 1    | 0.40 | N.A. | N.A. | 0.5  | 0.15 |
| Arsenic  | N.A. | 0.05 | 0.5  | 0.8  | 0.8  | 0.15 |
| Cadmium  | 1    | 0.15 | N.A. | N.A. | N.A. | 0.15 |
| Chromium | 1    | 0.05 | N.A. | 1    | 0.8  | 0.15 |
| Copper   | 1    | 0.05 | 0.05 | 1    | N.A. | 0.15 |
| Mercury  | 1    | N.A. | 0.80 | 0.9  | 0.2  | 0.50 |
| Lead     | 1    | 0.75 | 0.05 | 0.1  | N.A. | 0.15 |
| Zinc     | 1    | 0.15 | 0.05 | 0.1  | 0.8  | 0.15 |

[f] Uses not embodied in final products include catalysts, solvents, reagents, bleaches, etc. In some cases a chemical is embodied but there are some losses in processing. Losses in chemical manufacturing are included here. Major examples include copper and mercury catalysts (especially in chlorine manufacturing); copper, zinc, and chromium as mordants for dyes; mercury losses in felt manufacturing; chromium losses in tanning; lead in desulfurization of gasoline; and zinc in rayon spinning. In some cases, annual consumption is actually loss replacement and virtually all of the material is dissipated. Detonators such as mercury fulminate and lead azide (and explosives) are included in this category.

[g] Uses embodied in final products other than paints or batteries include fuel additives (e.g., tetraethyl lead), anticorrosion agents (e.g., zinc dithiophosphate), initiators and plasticizers for plastics (e.g., zinc oxide), wood preservatives, and chromium salts embodied in leather. Losses to the environment occur when the embodying product is used, for example, gasoline containing tetraethyl lead is burned and largely (75%) dispersed into the atmosphere. However, copper, chromium, and arsenic are used as wood preservatives and dispersed only if the wood is later burned or incinerated. In the case of silver (photographic film), it is assumed that 60% is later recovered.

[h] Includes agricultural pesticides, herbicides, and fungicides. Uses are dissipative, but heavy metals are largely immobilized by soil. Arsenic and mercury are exceptions because of their volatility.

[i] Biocides used in industrial, commercial, or residential applications. Loss rates are higher in some cases.

[j] Includes primarily pharmaceuticals (including cosmetics), germicides, etc., as well as dental filling material. Most uses are dissipated to the environment through wastewater. Silver and mercury dental fillings are likely to be buried with cadavers.

[k] Miscellaneous emissions not counted elsewhere.

SOURCE: Ayres et al. (1988).

effect") due to the rising level of carbon dioxide is still quantitatively uncertain, the qualitative impact is likely to be adverse.

Carbon monoxide is, of course, toxic to humans and has been implicated in health problems among urban populations. It is less well known that carbon monoxide plays an active, and not necessarily benign, role in a number of atmospheric chemical reactions before it is ultimately oxidized to carbon dioxide. Inefficient combustion processes convert up to 10 percent

TABLE 3 Emissions from Consumptive Uses: Heavy Metals, United States, 1980 (metric tons)

| Metal | Metallic Uses (Except Coatings and Electrical) | Protective Coverings | | Electrical | |
|---|---|---|---|---|---|
| | | Plating and Coating | Paints and Pigments | Batteries and Equipment | Other Electrical Uses, Instruments, etc. |
| Silver | 0.83 | a | 0.0 | 0.15 | 0.67 |
| Arsenic[b] | 0.04 | 0.0 | 0.0 | 1.97 | 0.0 |
| Cadmium | 0.04 | 136.1 | 116 | 7.81 | 0.0 |
| Chromium | 151.80 | 155.7 | 6,490 | 0.0 | 0.0 |
| Copper | 11,074.00 | 0.0 | 0.0 | 0.0 | 0.0 |
| Mercury[c] | 0.0 | 0.0 | 0.0 | 195.91 | 17.52 |
| Lead | 1,249.00 | 0.0 | 48,500 | 8,510.00 | 0.0 |
| Zinc[c] | 514.00 | 8,778.0 | 77,750 | 63.00 | 0.0 |

[a] Included in first column.
[b] 1979.
[c] 1977.

SOURCE: Ayres et al. (1988).

of fuel carbon into carbon monoxide; carbothermic reduction of iron ore and other metals generates even higher percentages. However, the average percentage over all processes is much smaller. Actual emissions of carbon monoxide to the atmosphere in the United States were about 110 million tons in 1970 (mostly from automobiles and trucks), with a carbon content of 47 million tons. Emission controls reduced the net output of carbon monoxide in 1980 to about 85 million tons (U.S. Environmental Protection Agency, 1986).

The discovery of chlorofluorocarbons (CFCs) in the stratosphere has raised an even more frightening prospect: ozone depletion (see Glas, this volume). In apparent confirmation of this phenomenon, an "ozone hole" has recently appeared in the stratosphere over Antarctica. This "hole" has reappeared each spring for several years and seems to be growing (Clark, 1987; Miller and Mintzer, 1986). Chlorofluorocarbons are chemically inert gases, discovered in 1928 and produced since the 1930s mainly as refrigerants and solvents and for "blowing" plastic foams (see Friedlander, this volume). The major use is for refrigeration and air conditioning. In most uses, CFCs are not released deliberately, but losses and leakage are inevitable.

The problems revealed so far may be only the beginning. If (perhaps it is better to say "when") the ozone level in the stratosphere is depleted, the effect will be to let more of the sun's ultraviolet radiation through to

| Chemical | | Biocidal Poison Uses | | | | |
|---|---|---|---|---|---|---|
| Industrial Catalysts, Reagents, Explosives, etc. | Consumer Uses, Additives, Extenders, Photography, etc. | Agricultural Pesticides, Herbicides, Fungicides | Nonagricultural Pesticides (Except Medical) | Medical, Dental, Pharmaceutical | Miscellaneous | Total |
| 7.6 | 48 | 0 | 0 | 1.51 | 0.45 | 60 |
| 492.0 | 0 | 2,950 | 5,901 | 19.70 | 0 | 9,364 |
| 0 | 29 | 0 | 0 | 0 | 0.98 | 290 |
| 1,297.0 | 3,890 | 0 | 1,038 | 0 | 2,141.00 | 11,659 |
| 4,222.0 | 0 | 1,560 | 0 | 0 | 0 | 15,452 |
| 412.1 | 0 | 16 | 236 | 8.36 | 6.96 | 893 |
| 0 | 0 | 0 | 0 | 0 | 1,329.00 | 56,900 |
| 2,508.0 | 18,622 | 188 | 251 | 1,003.00 | 0 | 109,670 |

the earth's surface. One likely impact on humans is a sharp increase in the incidence of skin cancer, especially among whites. The ecological impact on vulnerable plant or animal species cannot be estimated, at present, but could be severe.

Methane, oxides of nitrogen ($NO_x$), and sulfur oxides ($SO_x$) are other residuals that have been seriously implicated in climatic or ecological effects. All three are generated by fossil fuel combustion, as well as other industrial processes. Like carbon monoxide, carbon dioxide, and the CFCs, they can be considered as metabolic products of industry. Methane is released in natural gas pipelines, petroleum drilling, coal mining, and several kinds of intensive agriculture (especially rice cultivation or cattle and sheep farming).

Nitrogen oxides are also coproducts of combustion. In effect, at high temperatures, some of the atmospheric nitrogen is literally "burned." Nitrogen oxides are implicated in many atmospheric chemical reactions (including those that create smog) and eventually oxidize to nitric acid, which contributes to acid rain. In the stratosphere, where nitrogen oxides are decomposed by ultraviolet radiation, atomic nitrogen also contributes to the destruction of ozone. Finally, oxides of nitrogen are "greenhouse gases" that contribute to climatic warming (Miller and Mintzer, 1986).

Sulfur oxides are generated by the combustion of sulfur-containing fuels—especially bituminous coal—and by the smelting of sulfide ores. Most copper, lead, zinc, and nickel ores are of this kind. In principle, sulfur can be recovered for use from all these activities, and the recovery rate is rising. However, the costs of recovery, especially from coal-burning electric power plants, are still far higher than the market value of the potential products (e.g., dilute sulfuric acid). Hence, for the present, calcium sulfites and sulfates—as well as $SO_x$—constitute a waste residual that must be disposed of. This is also true of fly ash.[4]

These examples show that although today's industrialized economic system may be in rough equilibrium in terms of supply and demand relationships, it is far from equilibrium in thermodynamic terms. Enormous quantities of fossil fuels and high-quality minerals are extracted each year to drive the economic engine. The economic system is stable somewhat in the way a bicycle and its rider are stable: if forward motion stops, the system will collapse. Forward motion in the economic system is technological progress.

## EVOLUTION OF INDUSTRIAL PROCESSES

It is generally accepted by economists that the mechanism that normally drives the evolution of industrial processes is technological innovation. The primary incentive for taking risks appears to be economic. A new process that saves one link in the chain between raw materials and finished materials or final goods can usually be justified through savings in materials and energy inputs or capital requirements, if not both. Moreover, process technology is inherently easier to protect from piracy than product technology. As noted, most chemical products are intermediates used in the production of other chemicals. Thus, final products result from chains, or sequences, of processes with an overall energy conversion efficiency that is the arithmetic product of the conversion efficiencies at each stage.[5] If the typical chain has three steps, each of which has a conversion efficiency of 0.7, the overall conversion efficiency of the chain is about 0.34. A four-step chain would have an overall efficiency of around 0.24. That is, the available energy embodied in the final product might be somewhere between 25 and 35 percent of the available energy of the original feedstocks. Because primary feedstocks are essentially indistinguishable from fuels, efficiency improvements translate directly into cost savings.

Clearly, a powerful long-term strategy for improving overall effectiveness in production is the development of new processes to shorten these process chains, bypassing as many intermediates as possible. In other words, one would like to be able to produce final products such as polyethylene or synthetic rubber directly from first-tier intermediates or even from primary feedstocks such as ethane and propane.[6] Biological organisms differ from industrial organizations in that they are able to build complex molecules directly from elementary building blocks with relatively few intermediates. Thus, biotechnology offers a long-term prospect of radically higher production efficiencies and correspondingly lower costs (U.S. Office of Technology Assessment, 1982).

Another long-term strategy for increasing effectiveness is better use of by-products and wastes. When a process can be justified on the basis of the market for its primary product, by-product sales can be highly profitable. This positive motivation to seek new uses is compounded by the fact that, because of some of the environmental problems already noted, waste generation is being increasingly discouraged by environmental regulation.

Moreover, waste disposal is increasingly expensive, and the cost is becoming more uncertain. Firms that buried toxic chemical wastes many years ago in landfills (methods that were regarded as acceptable at the time) are sometimes finding themselves saddled retroactively with heavy costs of digging up the same wastes and disposing of them again in a safer manner. The chemical waste dump used by Hooker Chemical Company at Love Canal, near Niagara Falls, is one example. Meanwhile, suitable sites for disposal of hazardous solid or liquid wastes are becoming scarce.[7] Materials that are recovered and reused internally or embodied in marketable products that can be readily and effectively recycled are less likely to cause this kind of problem.[8]

The incentives for technological innovation in the area of waste reduction and disposal are not always operating, especially where there are massive market failures. In practice, the "unpaid" environmental damage costs have been deferred and many of them will have to be paid by later generations. But as these costs become larger and more visible, there will be growing political pressure to force the producers and users of fossil fuels (and other materials such as heavy metals) to pay the costs of abating the resulting environmental damages.[9] Notwithstanding resistance by energy users, it seems inevitable that in the long run these costs will be added to the prices of fuels and materials. This, in turn, will create significant economic opportunities for innovations in the area of "low-waste" technologies.

## EVOLUTIONARY ECONOMICS AND GAIA

One view of evolution—and probably the most common view among scientists—can be characterized as "the myopic drunkard's walk." The drunkard's walk is not exactly random, but it tends to follow the path of least resistance in the short run. If a mutation or an innovation offers short-term advantages, it will be adopted. In this view, there is no long-range force or tendency to approach a distant goal. There is no Aristotelian "final cause." Most scientists tend to regard more teleological theories, including Lovelock's Gaia hypothesis (Lovelock, 1988), with skepticism, because of their aroma of mysticism.

In the case of biological evolution, indeed, it was difficult for a long time to suggest a likely mechanism whereby short-term advantages to the individual could in some cases be overridden in favor of longer-term benefits to the community or the species. Even when such a mechanism is suggested (e.g., the kinship hypothesis), there may be difficulties in explaining how it could have become programmed into the genetic code. Yet, such behavior is a biological fact. The only real argument is whether a "strict" myopic test of the immediate utility of every mutation would have directed biological evolution along pathways significantly different from those that have actually occurred.

In the case of economic evolution, there is no difficulty in identifying possible mechanisms for selecting goal-oriented pathways. The decision makers are humans with the ability to look and plan ahead. The question here is, How far ahead do industrialists plan? The answer surely varies with individuals and circumstances, but it can be decades and is seldom less than a few years. Thus, long-term goal orientation in economic life is not particularly rare.

To relate this general point to the present context, one might postulate a long-run evolutionary imperative favoring industrial metabolic technologies that result in reduced extraction of virgin materials, reduced loss of waste materials, and increased recycling of useful materials. For convenience, one might refer to this overall trend—if it exists—as an imperative to reduce materials intensiveness, or dematerialization (see Herman et al., this volume). In general terms, the same long-term tendency can be observed in the biosphere, as noted in the Appendix.

It must be pointed out that short-term economic incentives do not necessarily point in the direction indicated. For example, market forces appear to favor product differentiation and specialization, but these trends increase the costs of repair and recycling. In poor countries, such as India, there is virtually no such thing as "junk." Any manufactured product, no matter how old or obsolete, is likely to be repaired or rebuilt and retained in service as long as physically possible. When it can no longer be repaired, it will be disassembled and useful parts will be separated for further use; the remainder will be hand-separated by material (stainless steel, iron, aluminum, copper, rubber, plastic, paper) and recycled in "backyard" operations. By contrast, in advanced countries, manufactured products are becoming more and more complex and correspondingly difficult to repair. This is particularly true of electronic devices such as printed circuit boards and cathode-ray tubes.

Moreover, as products are designed to be more reliable, so that repairs are no longer "normal" and expected, disassembly is becoming more difficult and in many cases is actively discouraged. In fact, for warranty reasons, critical subassemblies are often sealed and must be either returned to the factory or discarded. Finally, the complexity of products is often reflected by the increasing complexity of materials, which makes recycling inherently more difficult. To take one simple example, worn-out woolen suits and dresses were once routinely collected and recycled, mainly into coats and blankets. The once-profitable wool-recycling industry has virtually disappeared, because most new clothes are "blends" of natural and artificial fibers that cannot be reprocessed economically.

The point of the last two paragraphs is that short-term economic incentives and trends are often inconsistent with the postulated long-term imperative. One must, therefore, also postulate other counteracting incentives and mechanisms. Without going into detail, such incentives and mechanisms (if they exist) must grow out of social and even political responses to perceived environmental problems. It is political action, ultimately, that creates the incentive structure—fiscal, monetary, and tax policy—and the regulatory environment within which economic incentives drive entrepreneurial activity. Having said all this, I believe that the "dematerialization imperative" is alive and well.

The question I now raise is the following: Does this (hypothetical) evolutionary imperative toward reduced material intensiveness have any additional specific implications for future industrial processes? Can it be used as a basis for forecasting? I think the answer is a qualified yes.

To take one example, a future industrial metabolism deriving its energy ultimately from nuclear fission, nuclear fusion, or the sun itself, rather than from fossil fuels, would necessitate a completely different chemical energy carrier system. The most likely bulk energy carrier appears to be hydrogen. Unless biotechnology offers an attractive alternative, it appears likely that hydrogen would be obtained from electrolytic or thermal decomposition of water. Of the two processes, thermal decomposition is more direct and consequently, if the scale is large enough, more efficient.[10] Intermediates that have been suggested for thermal decomposition processes include a variety of compounds of iron, calcium, strontium, mercury, copper, chromium, manganese, and vanadium with bromine or chlorine. Many, if not most, of these compounds are considered hazardous. In consideration of the extremely large scale on which any such process is likely to have to be carried out, it is obviously of the utmost importance that process intermediates be recycled with extremely high efficiency (of the order of 99.9 percent). This is a severe challenge for chemical engineering and one that will probably be a major concern in future decades.

As another example, if we postulate much higher prices for carbon-based fuels (to reflect the environmental damage associated with their use), the problem of mobile power takes on a new dimension. Electric propulsion based on nuclear electricity (from "inherently safe" high-temperature gas-cooled reactors) is one possibility favored by many. But practical implementation on a large scale requires a technological breakthrough of unlikely magnitude in the area of compact electrical energy storage systems. Hydrogen has also been suggested as a possible substitute for liquid fuel (most likely for jet aircraft), but liquid fuels are far more convenient for most purposes.[11]

The evolution of industrial metabolism can also be addressed from the perspective of eliminating dissipative uses of toxic heavy metals (arsenic, cadmium, chromium, copper, lead, mercury, nickel, silver, and zinc) and of halogenated hydrocarbons. All of these materials have been implicated in environmental problems ranging from "Minimata disease" (mercury poisoning) to erosion of the ozone layer in the stratosphere. In the long run, it appears likely that all of them will have to be replaced or used only in applications permitting an extremely high degree of recycling. It is probably safe to say that the industrial metabolism of the next century will recycle many of the waste products that are produced in the largest quantities today, notably sulfur, fly ash, and lignin wastes from the paper industry. Lignin wastes may yet turn out to be a useful growth medium for single-cell organisms, providing high-protein supplements for food products. By the same token, inherently dissipative uses of biologically active materials will have to decrease as the mistakes of the past are rectified. In particular, the underpricing of environmental and exhaustible resources must be reduced or even (temporarily) reversed.

## APPLICATIONS OF MATERIALS-BALANCE PRINCIPLES

The materials-balance principle, a straightforward application of the first law of thermodynamics (widely used in the design of chemical engineering systems, for example), is a potentially valuable and underutilized tool for using economic data in environmental analysis. Frequently, a combination of input data (obtainable from economic statistics), together with technical process data available from engineering analysis, gives a more reliable estimate of waste residual outputs than direct measurements alone could be expected to do. This principle is particularly true when the pollutant of concern is produced in relatively small quantities and is emitted together with large amounts of combustion products or process wastewater.

A good example of the mass-balance principle has been taken from a study of environmental problems in the aluminum industry. One of the major environmental problems associated with aluminum smelting in the past was the emission of gaseous fluorides from the smelter. The source of fluorine is the electrolytes (molten cryolite and aluminum fluoride) used as a solvent for alumina in the electrolytic cell. An unavoidable side reaction in the cell breaks down these electrolytes and releases some of the fluorine at the anode. Exact "recipes" for the production of aluminum are known only by the aluminum companies. However, a materials-balance analysis for the year 1973 suggested that for each 100 kilograms of aluminum produced, 2.1 kilograms of cryolite and 3 kilograms of aluminum fluoride were consumed as inputs (Ayres et al., 1978). Based on these estimates and straightforward chemistry, the aluminum industry would have accounted for 40 percent of the known production of hydrofluoric acid in that year, which is consistent with both official (U.S. Bureau of the Census, 1960–1975) and unofficial estimates (Ayres et al., 1978). In the absence of fluorine recovery facilities, all of the fluorine consumed by the industry must have been emitted eventually to the environment. In other words, the total amount of cryolite and aluminum fluoride consumed by the aluminum industry was (and is) exclusively to replace fluorine losses.

It is interesting to note that the fluoride emissions calculated by using materials-balance principles were about twice as high as the Environmental Protection Agency's published estimates at the time. The latter were based on direct (but unreliable and difficult to verify) measurements taken at a few smelter sites. Assuming the production and use statistics for hydrofluoric acid were correct, one would have to believe that the indirect estimate based on materials-balance considerations was probably more reliable than the estimate based on partial and questionable direct measurements.

Another application of materials-balance methodology is in the reconstruction of historical emissions data. This is a problem of some importance to basic environmental science, because the cumulative impact of air or water pollution over long periods can be evaluated only in relation to a baseline of some sort. In this context it becomes important to know more about emissions in the past, when no measurements were made. To be sure, sediments and ice cores offer some help, but not enough. The picture can be clarified considerably, however, with the help of synthetic models using production and consumption data (which are often imperfect, but better known than emissions) together with engineering analysis of processes. Sometimes process information is not even needed, as when emissions are linked directly to inputs.

For example, fairly good historical estimates of $SO_x$ emissions, required to analyze the long-term impact of acid rain, among other things, can be reconstructed easily from historical statistics on coal consumption and copper, lead, and zinc smelting (e.g., Gschwandtner et al., 1983). This is because the sulfur content of coal and metal ores can be assumed to be the same in the past as it is today, and until recently all of that sulfur was emitted straight to the atmosphere. Reconstruction of $NO_x$ emissions data is slightly more complicated, but the approach is basically similar (Gschwandtner et al., 1983). More complex reconstructions of historical emissions data have been undertaken recently, e.g., for the Hudson-Raritan estuary (Ayres et al., 1988). Two examples of material process-product flows, taken from the above study, are presented in Figures 1 (cadmium) and 2 (chromium).

Another way of using the materials-balance approach is in the analysis of materials "cycles." The water cycle, the carbon cycle, and the nitrogen

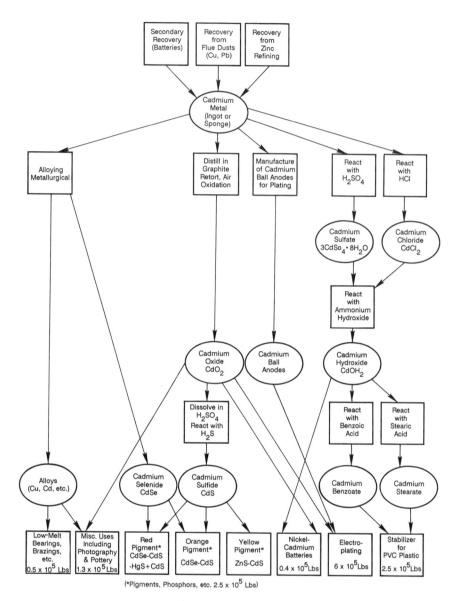

FIGURE 1 Cadmium process-product flows, circa 1969. Except for the bottom row, rectangles indicate processes; ellipses, products or reactions. SOURCE: Ayres et al. (1988).

cycle are familiar examples. The concept is also applicable, of course, to flows that are not truly cyclic, as in the case of arsenic (Figure 3). This concept has been widely used by geochemists, hydrologists, ecologists, and environmental scientists to organize and systematize their work. Such a presentation helps specify geographical scales of analysis. It also facilitates such comparisons as the relative importance of natural and anthropogenic sources. Finally, and potentially most important, it provides a starting point for detailed analysis of the effect of anthropogenic emissions on natural processes.

## CONCLUSION

It appears that we have methods to describe our industrial metabolism better, both qualitatively and quantitatively. Initial analyses reveal several important points, for example, that many materials uses are inherently dissipative and thus pose difficulties for recycling. Analysis also shows that although residuals do not disappear from the real world of human health and environmental quality, they do tend to disappear from the market domain. Thus, many environmental resources are underpriced and overused. It is also clear that where the production and use of by-products are concerned, many industrial processes involve multiple steps, resulting in a low level of system efficiency, especially in comparison with biological systems. The sum of the argument here suggests that we should not only postulate, but indeed endorse, a long-run imperative favoring an industrial metabolism that results in reduced extraction of virgin materials, reduced loss of waste materials, and increased recycling of useful ones.

## APPENDIX

### The Biosphere as a Materials-Transformation System

Three salient characteristics mark the difference between the naturally evolved biosphere and its human-designed industrial counterpart, the "synthesphere." The first is that the metabolic processes of biological organisms are derived (by photosynthesis) from a renewable source: sunlight. The second characteristic is that the metabolism of living organisms (cells) is executed by multistep regenerative chemical reactions in an aqueous medium at ambient temperatures and pressures.[12] Most process intermediates are regenerated within the cell. Reaction rates are controlled entirely by catalysts (enzymes). The energy transport function is performed in all living organisms by phosphate bonds, usually in the molecule adenosine triphosphate (ATP).[13]

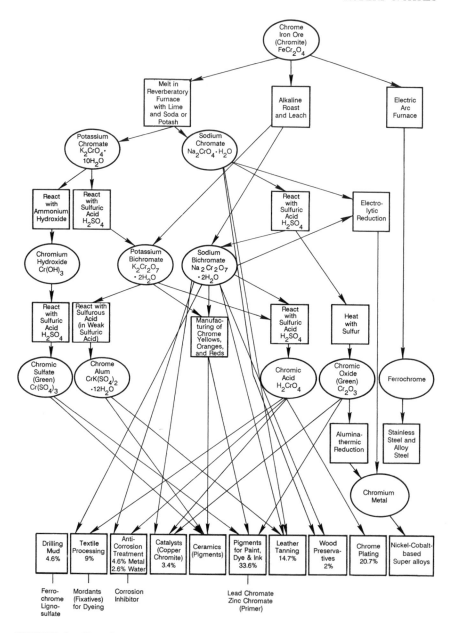

FIGURE 2 Chromium process-product flows, circa 1968. Except for the bottom row, rectangles indicate processes; ellipses, products or reactions. SOURCE: Ayres et al. (1988).

The third salient characteristic differentiating the biosphere from the industrial synthesphere is that, although individual organisms do generate process wastes—primarily oxygen in the case of plants and carbon dioxide and urea in the case of animals—the biosphere as a whole is extremely efficient at recycling the elements essential to life. Specialized organisms have evolved to capture nutrients in wastes (including dead organisms) and recycle them. A significant exception to this rule in the present geological epoch is the deposition of skeletal remains of zooplankton as sediments on the deep ocean floor. These remains are largely, but not entirely, calciferous.[14] Over geological time periods, some of this sedimentary material is likely to be recycled as chalk, limestone, or phosphate rock.

The biosphere as it now exists is a nearly perfect materials-recycling system, but this was not the case when life on earth began. The first and most critical evolutionary "invention," from which all else follows, was the process for replicating complex organic molecules. In effect, the information describing the entire living structure is stored as sequences of nucleic acids in the genetic substance known as deoxyribonucleic acid (DNA). The mechanism for storing, coding, transferring, and decoding that information apparently evolved some 4 billion years ago, before species differentiation. Both the code and the mechanism are common to all known living organisms.

The first cellular organisms, which appeared about 3.5 billion years ago, were prokaryotes (i.e., cells without nuclei). They obtained the energy needed to sustain the reproduction cycle from the anaerobic fermentation of organic molecules previously created by natural geophysical processes in an atmosphere containing no free oxygen. If the primitive atmosphere had contained oxygen, organic molecules could not have survived long enough to achieve the degree of complexity needed to construct self-reproducing systems.[15]

In cellular fermentation a molecule of glucose is split into two molecules of pyruvate. Energy is captured in the form of high-energy phosphate bonds in ADP (adenosine diphosphate) and ATP. Fermentation of a glucose molecule has a net yield of available energy in the form of two molecules of high-energy ATP, converted from the low-energy ADP form (Hinkle and McCarty, 1978). Further reactions in the cell convert the pyruvate to ethyl alcohol, lactic acid, and carbon dioxide, all of which are excreted as wastes. No oxygen is required. The fermentation-based forms of life could not have been the foundation for a sustainable ecosystem, however, because they were using up a finite stockpile of exploitable organic molecules.

The next great evolutionary innovation, about 3 billion years ago, was anaerobic photosynthesis. The first photosynthesizers were prokaryotic photobacteria. These organisms began to synthesize glucose from atmospheric carbon dioxide and sunlight, thus replacing the depleted organic "soup" of

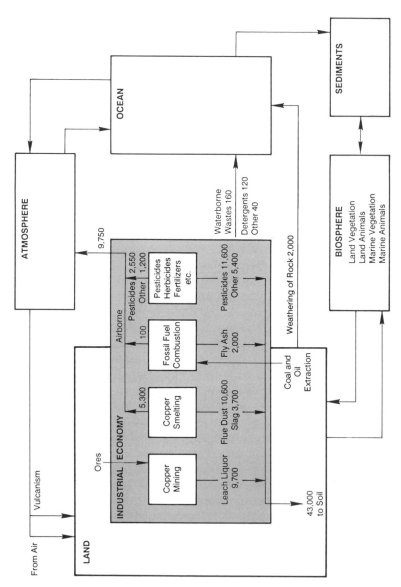

FIGURE 3 Simplified representation of arsenic pathways in the United States (metric tons), circa 1975. SOURCE: Ayres et al. (1988).

the primitive oceans. They also produced oxygen as a waste product. At first, the free oxygen was quickly removed by stromatolites, organisms that combined oxygen with iron dissolved in the oceans and precipitated as iron-rich reefs. This seems to have been the origin of hematite deposits being exploited as iron ores today. However, when the dissolved iron was used up, about 1.8 billion years ago, the oxygen level in the atmosphere began to rise, thereby increasing the rate of dissolution of all macromolecules. Thus, the ecosystem was still unsustainable, because it could neither tolerate nor recycle its own toxic wastes. The oxygen toleration problem was partially solved by a new class of aerobic photosynthesizers, the cyanobacteria (or blue-green algae). These appeared a little more than 2 billion years ago.

A third great evolutionary innovation—the substitution of respiration (using oxygen) for fermentation to obtain energy from organic molecules—solved the recycling problem. The respiration process begins the same way as fermentation, with the splitting of glucose into pyruvate (called glycolysis). However, in respiration, glycolysis is followed by a longer sequence of reactions, known collectively as the citric acid cycle. In effect, the pyruvates are oxidized enzymatically to carbon dioxide, with the formation of many more high-energy ATP molecules. In fact, each glucose molecule, when fully oxidized, yields 36 molecules of ATP, whereas the glycolysis stage alone yields only 2. The yield of available energy for further metabolic processes is, therefore, 18 times that of the fermentation process.

Because the respiration process is far more efficient than its precursor, aerobic respirators required much less organic material to sustain them. Thus, anaerobic organisms could not effectively compete with aerobic organisms in the presence of oxygen. (They still fill an environmental niche in sediments and deep oceans lacking free oxygen.)

With the "invention" of the citric acid cycle, the biosphere became sustainable within more self-defined system boundaries. Evolutionary developments since then—of which the most important were the development of the eukaryotes (cells with true nuclei) and the advent of sexual reproduction—made the system more diverse and more efficient. It is interesting to note that, despite the radical changes in energy metabolism which occurred, the basic scheme of macromolecular reproduction seems to have remained essentially unchanged for 4 billion years.

### *Industrial Transformation Processes*

In contrast to modern biological processes, industrial processes are almost exclusively energized by the combustion of fossil fuels, which (by definition) are not regenerated within the system. In this sense, the industrial system of today resembles the earliest stage of biological evolution, when the most primitive living organisms obtained their energy from a stock of organic molecules accumulated during prebiotic times. Instead of regenerative cycles powered by solar (or nuclear) energy, industrial processes are linear sequences of discrete, irreversible transformations (Figure 4). The sequence begins with extraction of raw materials, followed by physical separation and elimination of impurities, and subsequent reduction or recombination into convenient "first-tier" intermediates. This category includes primary metals and other elements in pure form, cellulose, sodium carbonate, ammonia, methane, ethane, propane, butane, benzene, xylene, methanol, ethanol, acetylene, ethylene, propylene, and some others. These materials, in turn, are subsequently recombined into desired chemical and physical forms.

Almost all of the processes for reducing metals from ores or producing first-tier intermediates are endothermic, that is, driven by externally supplied heat.[16] Processes often use catalysts, and rates and directions are fine tuned by controlled variation of temperatures, pressures, and flow rates or dwell times. There are five major categories of endothermic processes: (1) dehydration; (2) calcination; (3) "reducing" processes for splitting metal (or other) oxides into their constituents; (4) dehydrogenation processes, of which the simplest is the splitting of the water molecule;[17] and (5) processes for combining or synthesizing molecules that do not combine spontaneously at ambient temperatures or pressures. Energy is obtained initially from combustion (oxidation) and delivered either by process steam or by direct contact with the oxidation products. Comparatively few industrial processes are electrolytic: the production of aluminum, sodium, and chlorine and the refining of blister copper are the primary examples.

The most familiar example of dehydration is the production of plaster of paris from the mineral gypsum. Brick and ceramic manufacturing are also based to some extent on this process. The major example of calcination (and the origin of the name) is the production of calcium oxide (quicklime) from calcium carbonate (limestone) by driving off carbon dioxide. This is a major element in the manufacture of portland cement. Both dehydration and calcination are accomplished by the simple application of heat. The so-called carbothermic reduction process by which iron ore is converted to pig iron is a typical example of the third category.[18] In this reaction, coke is partially oxidized to carbon monoxide, which in turn reacts with the ore—at appropriate temperatures—to reduce the iron oxides, while converting the carbon monoxide to carbon dioxide.[19] The temperature determines which way the reaction goes. If the temperature rises too high (above 1300° C), any pure iron that is present will reoxidize.

Ammonia production exemplifies both the fourth and the fifth categories. Dehydrogenation (water splitting) is an example of a process by which synthesis gas for ammonia is produced. In a first stage of the reaction, steam reacts with natural gas, in the presence of a catalyst, to produce

# INDUSTRIAL METABOLISM

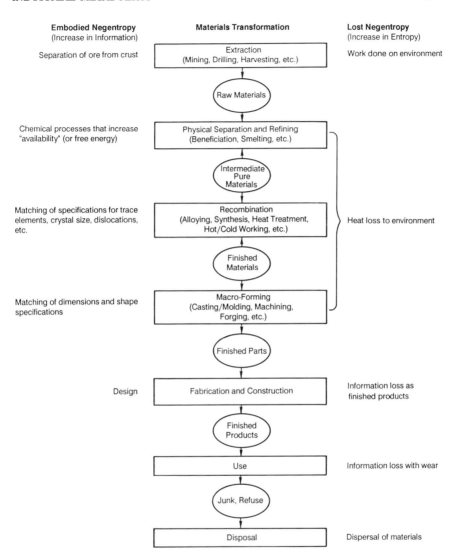

FIGURE 4 Representation of the economic system as a multistage system for the extraction, physical separation, recombination, formation, and consumption of materials. At the end of the sequence, materials are returned to the environment in a degraded form as waste. The process of creating a finished product (center column) concentrates information, or negentropy, in the product at the expense of increased entropy in the environment.

carbon monoxide and hydrogen. In a second stage, known as the water-gas shift reaction, carbon monoxide reacts with added steam to yield more hydrogen and carbon dioxide, which must be removed by a scrubber such as potassium carbonate. The hydrogen is then mixed with nitrogen gas (from the air) in a 3:1 ratio.[20] At very high pressures and temperatures, these ingredients combine endothermically—again in the presence of a catalyst—to form ammonia gas, the basis of virtually all nitrogen-containing compounds used by our industrial civilization.

Most of the reactions by which ammonia, chlorine, lime, sulfur, methanol, ethanol, acetylene, ethylene, propylene, or other first-tier intermediates are converted to other "downstream" compounds are exothermic and—in effect—self-energizing. For example, many second-tier intermediates are produced by controlled oxidation (e.g., sulfuric acid from sulfur, nitric acid from ammonia, acetaldehyde from ethanol or propane, acetic acid from acetaldehyde or butane, acetic anhydride from acetaldehyde, ethylene oxide from ethylene, propylene oxide from propylene, and so on). Most hydrogenation, chlorination, and hydrochlorination reactions are also exothermic, as are most reactions between strong acids and metals or hydroxides.

In effect, the first-tier intermediates are the energy carriers for subsequent reactions. They play a role somewhat analogous to that of ATP in biochemical systems. However, whereas the ATP is cyclically regenerated within the same cell, the first-tier intermediates are not regenerated but are physically embodied in downstream products. This is another fundamental difference between industrial metabolism in its present form and its biological analogue.

## NOTES

1. The three major German firms—Bayer, Hoechst, and BASF—were merged in the 1920s into a single giant under the name I.G. Farbenindustrie. Farben is the German word for "color." The name reflected their common origin as synthetic dye (color) manufacturers.
2. An important corollary is that the underpricing of environmental resources corresponds to an underpricing of those exhaustible mineral resources whose subsequent disposal as waste residuals causes harm to the environment. This is because of the lack of any link between the market price paid (for coal, oil, or whatever) and the subsequent cost of waste disposal or—more important—of uncompensated environmental or health damages such as bronchitis, asthma, emphysema, cancer, soil acidification, the greenhouse effect, and so on. Here the distinction between renewable and nonrenewable resources is critical: although renewable resources can obviously create pollution problems, such as sewage, they are almost invariably localized in nature and can be abated at moderate cost. This is emphatically not the case for combustion products of fossil fuels or dispersion of toxic heavy metals.
3. The total world output of carbon dioxide from fuels and cement manufacturing has been estimated to be 5.1 billion metric tons for 1982, of which 26.7 percent was attributable to North America (Marland and Rotty, 1984).

4. Fly ash is primarily a by-product of coal combustion. At present it is being recovered fairly efficiently, from stack gases of large utility boilers and industrial furnaces, by means of electrostatic precipitators. However, the ash itself has become a large-scale nuisance because there exists no use or market for it. The amounts are large: over 50 million tons are generated annually in the United States alone. Several possible remedies exist. Fly ash is a potential "ore" for several metals, especially iron, aluminum, and silicon. These metals could probably be recovered commercially if, for example, bauxite became unavailable (Ayres, 1982). Alternatively, fly ash could be used as a substitute, or more likely as a supplement, for portland cement in the manufacture of concrete and concretelike products. (Its major disadvantage in this application is that concrete made with fly ash does not harden and set as rapidly as the commercial variety. This has obvious economic costs, but so does the disposal of fly ash in landfills.) Another use of fly ash—already demonstrated in France—is as a means for the permanent disposal of toxic liquid wastes in the form of a hard, impermeable rocklike material suitable for long-term storage.
5. The "chain" analogy is an oversimplification, of course, because many processes yield more than one useful product (the chlor-alkali industry is an obvious example) and many products also require two or more inputs. Thus, the structure of the system as a whole is more like a "tree."
6. For example, the first process to manufacture acetylene proceeded by way of calcium carbide production (from coke and limestone) but was displaced by direct dehydrogenation of hydrocarbon feedstocks. The first production of acrylonitrile involved a reaction between ethylene oxide (itself the third step in a chain beginning with ethane) and hydrogen cyanide (made from ammonia). This was replaced by an acetylene cyanation process (acetylene being made from methane) and finally by a propylene ammoxidation process (propylene reacting directly with ammonia). Similarly, the first process in the manufacture of acetaldehyde started with acetylene (from calcium carbide) or ethanol (from ethylene). A newer process made acetaldehyde directly from ethylene. Thus, in the first example the calcium carbide stage was bypassed. In the second example the oxidation of ethylene was avoided, and in the third example the conversion of ethylene to alcohol was avoided.
7. The problem has been dramatized by several recent episodes in which cities have attempted to dispose of solid wastes by using private contractors who, in turn, thought to transport them to countries where disposal rules are nonexistent or weakly enforced. For example, the city of Philadelphia contracted with Joseph Paolino & Sons to dispose of incinerator ash. Paolino, in turn, contracted with Amalgamated Shipping, a Bahamian concern, to transport the ash to the Bahamas. However, the Bahamian government barred the dumping, and the ship carrying the ash was subsequently turned away from ports in the Dominican Republic, Haiti (after it had dumped 2,000 tons of ash), Honduras, Costa Rica, Guinea-Bissau, and the Cape Verde Islands. The ship apparently succeeded in dumping its load of ash somewhere in the Indian Ocean, after more than two years (*New York Times*, November 10, 1988).
8. Admittedly, it can still happen. Asbestos and polychlorinated biphenyls are two examples of materials that were once thought to be safe but have subsequently come to be regarded as hazardous and for which the major producers or users have had to spend billions of dollars to collect and dispose of them safely. Nevertheless, there would seem to be an economic opportunity for a "high-tech" resource-recovery firm to go into business reconverting fly ash and incinerator ash into its most valuable components, light metals such as aluminum, iron, potash, and titanium (Ayres, 1982).

A mineralized glassy residue of sodium-silica and heavy metals would, of course, remain for disposal (though it might also find uses as a construction material).
9. This is a straightforward implication of the widely accepted, but sporadically enforced, "polluter must pay" principle.
10. A number of possible processes for large-scale thermal decomposition of water have been suggested and studied in some detail, e.g., by the European Atomic Energy Commission (EURATOM). See, for example, Marchetti (1973).
11. Ammonia, hydrazine, and other compounds have also been suggested, but there is no clear leader at this time. A new study of the economic and technical feasibility of non-carbon-based liquid fuels would be helpful in clarifying the choices.
12. It is true that some organisms have evolved to function in the deep oceans under conditions of high pressure and salinity; others have evolved to function in surface waters, saline or fresh; still others have evolved to function in environments with very little water. Nevertheless, the internal environment of every cell is aqueous and the pressure inside each cell is essentially the same as the external pressure of the environment in which the organism lives.
13. For a concise summary of the biochemistry of energy transport in cells, see Schopf (1978).
14. The so-called manganese nodules, which are accretions of iron, manganese, copper, cobalt, and other transition elements, are evidently the result of some combination (as yet imperfectly understood) of biological, chemical, and geological processes (see, for example, Morgenstein, 1973).
15. Even in the absence of oxygen, it would seem that dissolution of macromolecules should proceed faster than synthesis (Wald, 1954). The exact evolutionary mechanism leading to self-reproduction is still obscure. Fixation of nitrogen must have been accomplished during this early period, because the early atmosphere was nearly transparent to ultraviolet radiation and any free ammonia in the atmosphere would have been quickly destroyed.
16. The obvious exceptions are elements occurring naturally (such as sulfur) and hydrocarbons that can be obtained by physical separation from natural gas (methane, ethane, propane, butane) or coal tar (benzene, xylene, toluene). Cellulose occurs naturally in some very pure forms (e.g., cotton), but it is usually obtained from wood pulp by a chemical digestion process.
17. Other major examples of dehydrogenation include the splitting of methane, ethane, propane, and butane to produce acetylene, ethylene, propylene, butylene, and butadiene.
18. Pig iron is a solution of iron carbide in iron, with a typical carbon content of approximately 6 percent. The conversion of pig iron to pure (wrought) iron or steel requires removing this carbon and then adding any desired alloying elements.
19. Multistage reactions such as the reduction of iron ore and the synthesis of ammonia, involving an intermediate (carbon monoxide) that is produced by the reaction and later consumed, are quite common in industry. Less common are reactions involving an intermediate that is not produced within the process but is recycled. One of the first examples of such a process was the Solvay (ammonia-soda) process for manufacturing synthetic sodium carbonate from sodium chloride and calcium carbonate. In this process, ammonium hydroxide reacts with calcium carbonate to yield ammonium carbonate and calcium hydroxide. Ammonium carbonate is converted to ammonium bicarbonate. When this reacts with sodium chloride, sodium carbonate and ammonium chloride are produced. Finally, calcium hydroxide and ammonium chloride are reacted to recover ammonia (gas) for recycling and calcium chloride. The latter is a low-value by-product.

20. In principle, the nitrogen gas and steam could be produced together by partial oxidation of natural gas (or any hydrocarbon) in air to yield a mixture of nitrogen, steam, and carbon monoxide. The steam in the hot combustion products could then be reacted with additional natural gas to generate hydrogen and more carbon monoxide to be used as feedstock for the shift reaction. However, the presence of nitrogen before it is wanted complicates the engineering unreasonably.

# REFERENCES

Ayres, R. U. 1978. Resources, Environment and Economics: Applications of the Materials/Energy Balance Principle. New York: John Wiley & Sons.

Ayres, R. U. 1982. Coalplex: An integrated energy/resources system concept. United Nations Environment Program Seminar on Environmental Aspects of Technology Assessment, United Nations, Geneva, November–December 1980.

Ayres, R. U., J. Cummings-Saxton, and E. Weinstein. 1978. Assessment of methodologies for indirect impact assessment. Final Report (IRT-468-R/a) prepared for U.S. Environmental Protection Agency. Washington, D.C.: International Research and Technology Corporation.

Ayres, R. U., L. W. Ayres, J. A. Tarr, and R. C. Widgery. 1988. An historical reconstruction of major pollutant levels in the Hudson-Raritan Basin: 1880–1980. National Oceanic and Atmospheric Administration Technical Memorandum NOS OMA 43, 3 vols. Rockville, Md.

Clark, R., ed. 1987. The Ozone Layer [Series: UNEP/GEMS Environment Library], Vol. 2. United Nations Environment Program, Nairobi.

Gschwandtner, G., K. C. Gschwandtner, and K. Eldridge. 1983. Historic Emissions of Sulfur and Nitrogen Oxides in the U.S. 1900–1980. Report to U.S. Environmental Protection Agency. Durham, N.C.: Pacific Environmental Services, Inc.

Hinkle, P. C., and R. E. McCarty. 1978. How cells make ATP: The prevailing theory is the "chemiosomotic" one. Scientific American 238(3):104–123.

Lovelock, J. 1988. The Ages of Gaia: A Biography of Our Living Earth. New York: Norton.

Marchetti, C. 1973. Hydrogen and energy. Chemical Economy and Engineering Review 5(1):7–25.

Marland, G., and R. M. Rotty. 1984. Carbon dioxide emissions from fossil fuels: 1950–1982. Tellus 36B(4):232–261.

Miller, A. S., and I. M. Mintzer. 1986. The Sky is the Limit: Strategies for Protecting the Ozone Layer, Research Report (3). Washington, D.C.: World Resources Institute.

Morgenstein, M., ed. 1973. Papers on the Origin and Distribution of Manganese Nodules in the Pacific and Propects for Exploration. Symposium, Hawaii Institute of Geophysics, Honolulu.

Multhauf, R. P. 1967. Industrial chemistry in the nineteenth century. Pp. 468–488 in Technology in Western Civilization: The Emergence of Modern Industrial Society—Earliest Times to 1900, M. Kranzberg and C. W. Pursell, Jr., eds. New York: Oxford University Press.

New York Times. November 10, 1988. After two years at sea, ship dumps U.S. ash. Vol. 138, 10(N), c4(L).

Schopf, J. W. 1978. The evolution of the earliest cells. Scientific American 239(3):110–138.

U.S. Bureau of the Census. 1960–1975. Statistical Abstract of the United States. Washington, D.C.: U.S. Government Printing Office.

U.S. Environmental Protection Agency. 1986. National Air Pollution Emission Estimates, 1940–1984 (EPA-450/4-85-014). Research Triangle Park, N.C.: Office of Air Quality Planning and Standards, U.S. Environmental Protection Agency.

U.S. Office of Technology Assessment. 1982. Genetic Technology: A New Frontier. Boulder, Colo.: Westview Press.

Wald, G. 1954. The origin of life. Scientific American 191(August):45–53.

# Dematerialization

ROBERT HERMAN, SIAMAK A. ARDEKANI, AND
JESSE H. AUSUBEL

Until recently the role of consumption as a driving force for environmental change has not been widely explored. This may be due in part to the difficulty of collecting suitable data. The present chapter approaches the consumption of materials from the perspective of the forces for materialization or dematerialization of industrial products beyond the underlying and obviously very powerful forces of economic and population growth. Examination can occur on both the unit and the aggregate level of materials consumption. Such study may make it possible to assess current streams of materials use and, based on environmental implications, may suggest directions for future materials policy.

The word *dematerialization* is often broadly used to characterize the decline over time in weight of the materials used in industrial end products. One may also speak of dematerialization in terms of the decline in "embedded energy" in industrial products. Colombo (1988) has speculated that dematerialization is the logical outcome of an advanced economy in which material needs are substantially satiated.[1] Williams et al. (1987) have explored relationships between materials use and affluence in the United States. Perhaps we should first ask the question: Is dematerialization taking place? The answer depends, above all, on how dematerialization is defined. The question is particularly of interest from an environmental point of view, because the use of less material could translate into smaller quantities of waste generated at both the production and the consumption phases of the economic process.

But less is not necessarily less from an environmental point of view. If smaller and lighter products are also inferior in quality, then more units would be produced, and the net result could be a greater amount of waste generated in both production and consumption. From an environmental viewpoint, therefore, (de)materialization should perhaps be defined as the change in the amount of waste generated per unit of industrial products. On the basis of such a definition, and taking into account overall production and consumption, we have attempted to examine the question of whether dematerialization is occurring. Our goal is not to answer definitively the question whether society is dematerializing but rather to establish a framework for analysis to address this overall question and to indicate some of the interesting and useful directions for study. We have examined a number of examples even though the data are not complete.

Undoubtedly, many industrial products have become lighter and smaller with time. Cars, dwelling units, television sets, clothes pressing irons, and calculators are but a few examples. There is, of course, usually a lower bound regarding how small objects such as appliances can be made and still be compatible with the physical dimensions and limitations of human beings (who are themselves becoming larger), as well as with the tasks to be performed.[2] Apart from such boundary conditions on size and possibly weight of many industrial product units, dematerialization of units of products is perceived to be occurring.

An important question is how far one could drive dematerialization. For example, for the automobile, how is real world safety related to its mass? In a recent study, Evans (1985) found that, given a single-car crash, the unbelted driver of a car weighing about 2,000 pounds is about 2.6 times as likely to be killed as is the unbelted driver of an approximately 4,000-pound car. The relative disadvantage of the smaller car is essentially the same when the corresponding comparison is made for belted drivers. For two-car crashes it was found that the driver of a 2,000-pound car crashing into another 2,000-pound car is about 2.0 times as likely to be injured seriously or fatally as is the driver of a 4,000-pound car crashing into another 4,000-pound car. These results suggest one of the reasons that dematerialization by itself will not be a sufficient criterion for social choice about product design. If the product cannot be practically or safely reduced beyond a certain point, can the service provided by the product be provided in a way that demands less material? To return to the case of transportation, substituting telecommunications for transportation might be a dematerializer, but we have no data on the relative materials demand for the communications infrastructure versus the transportation infrastructure to meet a given need. In any case, demands for communication and transportation appear to increase in tandem, as complementary goods rather than as substitutes for one another.

It is interesting to inquire into dematerialization in the world of miniaturization, not only the world of large objects. In the computer industry, for example, silicon wafers are increasing in size to reduce material losses in cutting. This is understandable if one considers that approximately 400 acres of silicon wafer material are used per year by IBM Corporation at a cost of about $100 million per acre. A processed wafer costs approximately $800, and the increase in total wafer area per year is about 10–15 percent. Although silicon wafers do not present a waste disposal problem from the point of view of volume, they are environmentally important because their manufacture involves the handling of hazardous chemicals. They are also interesting as an example of how the production volume of an aggressive new technology tends to grow because of popularity in the market. Moreover, many rather large plastic and metal boxes are required to enclose and keep cool the microchips made with the wafers, even as the world's entire annual chip production might compactly fit inside one 747 jumbo jet. Thus, such new industries may tend to be simultaneously both friends and foes of dematerialization.

The production of smaller and lighter toasters, irons, television sets, and other devices in some instances may result in lower-quality products and an increased consumer attitude to "replace rather than repair." In these instances, the number of units produced may have increased. Although dematerialization may be the case on a per-unit basis, the increasing number of units produced can cause an overall trend toward materialization with time. As an example, the apparent consumption of shoes, which seem increasingly difficult to repair, has risen markedly in the United States since the 1970s, with about 1.1 billion pairs of nonrubber shoes purchased in 1985, compared with 730 million pairs as recently as 1981 (Table 1).

In contrast, improvements in quality generally result in dematerialization, as has been the case for tires. The total tire production in the United States has risen over time (Figure 1), following from general increases in both the number of registered vehicles and the total miles of travel. However, the number of tires per million vehicle miles of travel has declined (Figure 2). Such a decline in tire wear can be attributed to improved tire quality, which results directly in a decrease in the quantity of solid waste due to discarded tires. For example, a tire designed to have a service life of 100,000 miles could reduce solid waste from tires by 60–75 percent (Westerman, 1978). Other effective tire waste reduction strategies include tire retreading and recycling, as well as the use of discarded tires as vulcanized rubber particles in roadway asphalt mixes.

Dematerialization of unit products affects, and is influenced by, a number of factors besides product quality. These include ease of manufacturing, production cost, size and complexity of the product, whether the product is to be repaired or replaced, and the amount of waste to

TABLE 1 Apparent Consumption of Nonrubber Shoes in the United States

| Year | Total Consumption (million pairs) | Shoes per Capita per Year |
|---|---|---|
| 1970 | 802 | 3.9 |
| 1975 | 728 | 3.4 |
| 1976 | 786 | 3.6 |
| 1977 | 746 | 3.4 |
| 1978 | 788 | 3.6 |
| 1979 | 833 | 3.7 |
| 1980 | 745 | 3.3 |
| 1981 | 730 | 3.2 |
| 1982 | 830 | 3.6 |
| 1983 | 912 | 3.9 |
| 1984 | 1,018 | 4.3 |
| 1985 | 1,097 | 4.6 |

SOURCE: U.S. Bureau of the Census (1975-1985).

be generated and processed. These factors influence one another as well (Figure 3). For example, the ease of manufacture of a particular product in smaller and lighter units may result in lower production cost and cheaper products of lower quality, which will be replaced rather than repaired on breaking down. Although a smaller amount of waste will be generated on a per-unit basis, more units will be produced and disposed of, and there may be an overall increase in waste generation at both the production and the consumption ends.

Another factor of interest on the production end is scale. One would expect so-called economies of scale in production to lead to a set of facilities that embody less material for a given output. Does having fewer, larger plants in fact involve significantly less use of material (or space) than having more, smaller ones? At the level of the individual product, the shift from mainframe computers to personal computers, driven by desires for local independence and convenience, may also be in the direction of materialization.

Among socioeconomic factors influencing society's demand for materials are the nature of various activities, composition of the work force, and income levels. For example, as a predominantly agricultural society evolves toward industrialization, demand for materials increases, whereas the transition from an industrial to a service society might bring about a decline in the use of materials. Within a given culture, to what extent are materials use and waste generation increasing functions of income?

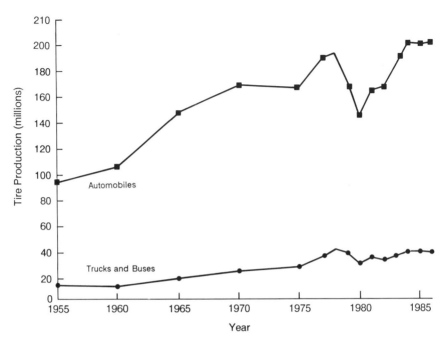

FIGURE 1 Production of automobile, truck, and bus tires in the United States. SOURCE: U.S. Bureau of the Census (1975–1985).
NOTE: Lines connecting data points are for clarity only.

The spatial dispersion of population is a potential materializer. Migration from urban to suburban areas, often driven by affluence, requires more roads, more single-unit dwellings, and more automobiles with a consequent significant expansion in the use of materials. The movement from large, extended families sharing one dwelling to smaller, nuclear families may be regarded as a materializer if every household unit occupies a separate dwelling. Factors such as photocopying, photography, advertising, poor quality, high cost of repair, and wealth generally force materialization. Technological innovation, especially product innovation, may also tend to force materialization, at least in the short run. For example, microwave ovens, which are smaller than old-fashioned ovens, have now been acquired by most American households. However, they have come largely as an addition to, not a substitute for, previous cooking appliances. In the long term, if microwave ovens truly replace older ovens, this innovation may come to be regarded as a dematerializer. National security and war, styles and fashions, and fads may also function as materializers by accelerating production and consumption. Demand for health and fitness, local mobility, and travel may spur materialization in other ways.

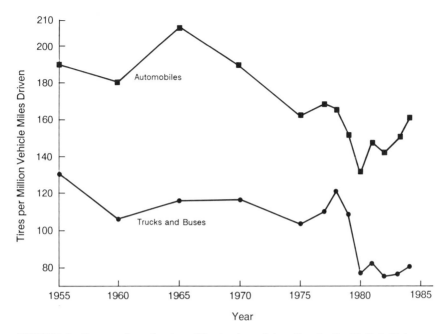

FIGURE 2 Consumption of automobile, truck, and bus tires in the United States per million vehicle miles driven. SOURCE: U.S. Bureau of the Census (1975–1985).
NOTE: Lines connecting data points are for clarity only.

The societal driving forces behind dematerialization are, at best, diverse and contradictory. However, the result may indeed be a clear trend in materialization or dematerialization. This could be determined only through collection and analysis of data on the use of basic materials with time, particularly for industry and especially for products with the greatest materials demand. Basic materials such as metals and alloys (e.g., steel, copper, aluminum), cement, sand, gravel, wood, paper, glass, ceramics, and rubber are among the materials that should be considered. The major products and associated industries that would be interesting to study could well include roads, buildings, automobiles, appliances, pipes (metal, clay, plastic), wires, clothing, newsprint and books, packaging materials, pottery, canned food, and bottled or canned drinks.

Hibbard (1986) reported without much detail that annual per capita intensity of materials use in the United States remained nearly constant between 1974 and 1985 at about 20,000 pounds. It would be useful to confirm this finding and extend the data to explore the extent to which such a fact might result from changes in gross national product (GNP), materials substitutions, market saturation, or other factors of the kind mentioned above. About two-thirds of Hibbard's estimate comes from

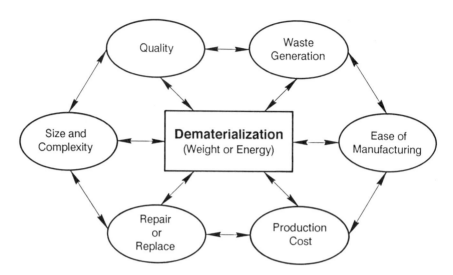

FIGURE 3 Factors affecting, and affected by, the dematerialization process. Economic and population growth, of course, also strongly interact with many of the factors.

stone, and from sand and gravel for concrete. These materials may be less important for environmental quality than others that are more active in our "industrial metabolism" (see Ayres, this volume). Further thought should be given to defining baskets of materials whose use over time might form the most meaningful indicators with respect to environment from the point of view of processing and disposal.

Table 2 shows the consumption of carbon steel as a function of time across various major end products. As can be seen, the use of steel in two major industrial activities, namely, construction and automobile manufacture, clearly has been in decline. This significant dematerialization trend has come about by virtue of the use of lightweight, high-strength alloys, and synthetics as substitutes for steel and cast iron. The trend is especially evident in the automobile industry where large weight and size reductions were achieved by materials substitutions in the 1970s in order to conserve energy. Table 3, the estimated pounds of materials used in a typical U.S.-manufactured car, shows that the use of plain carbon steel declined by 475 pounds per car in the 10 years examined, 1978–1988. On the other hand, the use of high-strength steel, plastic composites, and aluminum increased by 99, 43, and 36.5 pounds, respectively, in the same period. The result is a total reduction in weight of a typical U.S. car of about 400 pounds from 1978 to 1988. In the construction industry, however, caution must be exercised in associating the decline in steel use with dematerialization, because such a decline could be indicative of the increased popularity of concrete over steel as the basic construction material for aesthetic, technical, or cost reasons.

Growth in the use of advanced materials is expected to continue. For example, it is anticipated that by 1997 the world market for fabricated advanced polymer composites will be almost triple its 1987 level (Miller, 1988, pp. 57–59). These changes will significantly affect the industries producing conventional materials because the automotive industry has traditionally been a major consumer of these materials. In 1978, for example, the automotive industry used 22 percent of the total U.S. steel consumption and 17 percent of aluminum consumption (Motor Vehicle Manufacturers Association, 1982, p. 6).

The significant decline in the use of steel in the automobile industry provides strong evidence in support of dematerialization at the production end. An examination of energy consumption in selected national economies between 1973 and 1985 further underscores an industrial trend in efficiency and dematerialization (Table 4). Although total energy consumption in most countries increased considerably during this period, the energy consumed per 1980 constant GNP dollar declined in 9 out of 10 nations examined. This result may be explained in part by energy efficiency in production or by an increasing GNP associated with the services sector. Whether increasing energy efficiency is a net dematerializer is not clear. Often, increasing energy efficiency involves substituting durable capital goods in the form of better or larger amounts of building materials such as insulation. However, further evidence of dematerialization at the production end is provided by data on industrial solid waste generation, which show a significant decline from 1979 to 1982 (Figure 4).

The generation of municipal solid waste, also shown in Figure 4, has been on the increase. Examination of trends in municipal solid waste generation (total and by each component) provides insight into materialization and dematerialization at the consumption end. The data on municipal solid waste generation suggest a trend toward materialization at the consumption end. Table 5, for example, shows the total amount of paper waste in the municipal solid waste in the United States. As can be seen, from 1960 to 1982 there was an approximately 75 percent increase in the total paper waste generated, as well as an approximately 35 percent increase in the paper disposed per capita. Such increases are to be viewed in light of predictions that the advent of computers would reduce the use and wastage of paper. A possible contribution to this rise in paper waste is the increase in circulation of daily newspapers from a total of 53.8 million in 1950 to 62.8 million in 1985 (U.S. Bureau of the Census, 1975–1985).

However, other factors must be taken into account, such as changes in the average size and number of pages of newspapers over the country, as well as the amount of wastepaper that is used by the industry for printing

TABLE 2  Apparent Consumption of Carbon Steel Products in the United States by End Use, 1970-1982 (million tons)

| Year | Motor Vehicles | Construction | Consumer Durables | Producer Durables | Rail Transportation | Containers and Packaging | Oil and Gas Industry | Other | Total |
|---|---|---|---|---|---|---|---|---|---|
| 1970 | 19.5 | 22.9 | 5.6 | 11.6 | 3.4 | 9.0 | 1.4 | 16.8 | 90.3 |
| 1971 | 23.8 | 22.7 | 5.7 | 11.5 | 3.4 | 8.4 | 1.6 | 17.7 | 94.8 |
| 1972 | 24.9 | 22.3 | 6.1 | 12.7 | 3.1 | 8.0 | 1.7 | 18.7 | 97.4 |
| 1973 | 29.8 | 26.4 | 6.5 | 14.3 | 3.4 | 9.2 | 2.3 | 20.9 | 112.8 |
| 1974 | 24.6 | 27.7 | 6.1 | 14.6 | 3.8 | 9.6 | 2.8 | 20.9 | 110.0 |
| 1975 | 19.3 | 18.3 | 4.1 | 10.8 | 3.3 | 7.0 | 3.0 | 14.8 | 80.7 |
| 1976 | 26.8 | 19.0 | 5.1 | 12.2 | 3.2 | 8.0 | 2.0 | 16.0 | 92.3 |
| 1977 | 28.5 | 20.1 | 5.7 | 12.9 | 3.6 | 8.0 | 3.0 | 16.8 | 98.6 |
| 1978 | 27.7 | 22.5 | 5.7 | 14.2 | 3.7 | 7.9 | 3.2 | 20.3 | 105.4 |
| 1979 | 22.0 | 23.0 | 5.8 | 14.2 | 4.3 | 8.0 | 3.0 | 21.9 | 102.2 |
| 1980 | 14.5 | 18.9 | 4.5 | 11.7 | 3.3 | 6.7 | 4.1 | 21.4 | 85.1 |
| 1981 | 15.8 | 19.2 | 4.8 | 12.2 | 3.2 | 6.5 | 5.1 | 24.9 | 91.7 |
| 1982 | 10.6 | 14.7 | 4.0 | 8.4 | 3.0 | 5.1 | 3.1 | 20.7 | 69.6 |

NOTE: These data were constructed by aggregating various American Iron and Steel Institute categories and by allocating shipments to service centers and imports to the end-use sectors.

SOURCE: National Academy of Engineering (1985).

TABLE 3  Estimated Material in a Typical U.S. Car (pounds)

| Material | 1978 | 1984 | 1986 | 1988 |
|---|---|---|---|---|
| Plain carbon steel | 1,915.0 | 1,526.0 | 1,470.0 | 1,440.0 |
| High-strength steel | 133.0 | 210.0 | 223.5 | 232.0 |
| Stainless steel | 26.0 | 28.5 | 30.5 | 31.0 |
| Other steels | 55.0 | 54.0 | 55.5 | 45.0 |
| Iron | 512.0 | 481.0 | 465.5 | 457.0 |
| Plastics/plastic composites | 180.0 | 204.0 | 216.0 | 223.0 |
| Fluids and lubricants | 198.0 | 189.0 | 181.0 | 178.0 |
| Rubber | 146.5 | 138.0 | 134.5 | 134.0 |
| Aluminum | 112.5 | 136.5 | 139.5 | 149.0 |
| Glass | 86.5 | 85.5 | 85.5 | 85.0 |
| Copper fabrications and electrical components | 37.0 | 43.5 | 46.0 | 49.0 |
| Zinc die castings | 31.0 | 17.5 | 18.0 | 19.5 |
| Other materials | 137.0 | 118.5 | 105.0 | 124.5 |
| Total | 3,569.5 | 3,232.0 | 3,170.5 | 3,167.0 |

NOTE: Estimates are based on U.S. models only, including family vans and wagons.

SOURCE: Stark (1988, pp. 33-37).

TABLE 4  Energy Intensity of Selected National Economies, 1973-1985

| Country | Energy (megajoules) per 1980 Dollar of GNP | | | | Change, 1973-1985 (percent) |
|---|---|---|---|---|---|
| | 1973 | 1979 | 1983 | 1985 | |
| Australia | 21.6 | 23.0 | 22.1 | 20.3 | -6 |
| Canada | 38.3 | 38.8 | 36.5 | 36.0 | -6 |
| Federal Republic of Germany | 17.1 | 16.2 | 14.0 | 14.0 | -18 |
| Greece[a] | 17.1 | 18.5 | 18.9 | 19.8 | +16 |
| Italy | 18.5 | 17.1 | 15.3 | 14.9 | -19 |
| Japan | 18.9 | 16.7 | 13.5 | 13.1 | -31 |
| Netherlands | 19.8 | 18.9 | 15.8 | 16.2 | -18 |
| Turkey | 28.4 | 24.2 | 25.7 | 25.2 | -11 |
| United Kingdom | 19.8 | 18.0 | 15.8 | 15.8 | -20 |
| United States | 35.6 | 32.9 | 28.8 | 27.5 | -23 |

[a]Energy intensity increased as a result of a move toward energy-intensive industries such as metal processing.

SOURCE: International Energy Agency (1987).

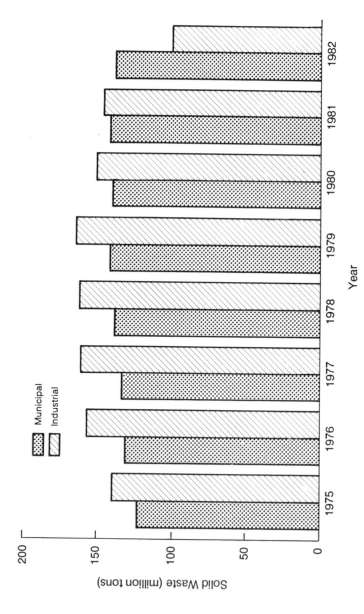

FIGURE 4  Disposal of municipal and industrial solid waste in the United States. SOURCE: U.S. Bureau of the Census (1975–1985).

TABLE 5 Amount of Paper in the Total Municipal Solid Waste Generated in the United States

| Year | Paper Wasted (million tons) | Paper Wasted (pounds per capita per day) | Paper Recycled (million tons) |
|---|---|---|---|
| 1960 | 29.9 | 0.91 | 5.4 |
| 1965 | 37.9 | 1.07 | 5.7 |
| 1970 | 40.4 | 1.08 | 6.8 |
| 1975 | 42.7 | 1.08 | 8.2 |
| 1976 | 49.1 | 1.24 | 9.7 |
| 1977 | 50.8 | 1.27 | 10.5 |
| 1978 | 53.4 | 1.32 | 10.6 |
| 1979 | 55.5 | 1.35 | 11.6 |
| 1980 | 54.1 | 1.30 | 11.8 |
| 1981 | 55.5 | 1.32 | 11.4 |
| 1982 | 52.2 | 1.23 | 10.8 |

SOURCE: U.S. Bureau of the Census (1975-1985).

new newspapers. Wastepaper consumption in the newspaper industry rose from about 2.6 million short tons in 1977 to 3.6 million short tons in 1987 (Institute of Scrap Recycling Industries, 1988, p. 22).

Recently, there has been a great deal of interest in the paradox associated with the proliferation of paper in our sociotechnical culture. The following discussion on this point is based on a recent article by Tenner (1988). We were all encouraged in the past to believe that information technology, as a by-product, was going to reduce the consumption of paper significantly. As we all now know, the reverse has transpired, with paper prices rising and trees in jeopardy. Consumption in the United States of writing and printing paper increased in 1959–1986 from about 7 to 22 million tons, and in the short period 1981–1984 the use of paper by U.S. businesses rose from 850 billion to 1.4 trillion pages. It is estimated that between 1986 and 1990, printed material may increase from about 2.5 to 4 trillion pages. In 1988 newsprint production was approaching capacity at about 12 million metric tons, and in the Pan Am Building in New York City a newsstand is reported to carry more than 2,000 magazines!

Banks have rid us of the savings account passbook, but in its place there is a spate of paper. Consumers have resisted reliance on home computer on-line services. Moreover, attempts by banks not to provide customers with canceled checks have failed; in 1985 U.S. banks processed some 45 billion checks. Plastic credit cards generate considerable amounts of paper, as do automated teller machines. The Rush Medical Library, Chicago, used about 188 linear miles of paper in its photocopy machines in the year 1982–1983, and the Princeton University computer center used close to 6 million pages of letter-sized laser paper in 1986, plus about 4,500 cartons of impact printout paper. Harvard's computer printers use more than 22 million pages a year, not including personal and faculty computers.

The question has been asked, What was wrong with the assumption that electronics would substitute for paper? Apparently nobody anticipated that the microchip would catalyze the burgeoning of paper to such an enormous extent. It would appear that the information age technicians did not understand that the amount of information was not fixed and that electronic information was not simply a substitute for paper. Computers are storing greater quantities of more kinds of information than ever before in extremely compact form, but people prefer reading from the printed page rather than the average computer screen, which in order to have excellent resolution must be improved by a factor of about 10. In addition, there is an increase in office workers compared to those in manufacturing jobs, and this shift leads to an increase in precisely the kind of people who generate paper. Note also that it is easy to produce photocopies compared to the old days, when making carbon copies was indeed a great burden.

In 1959 when Xerox introduced its dry copier, a consulting company estimated that no more than 5,000 such copiers would be required in the entire United States. The huge mailings today from businesses and various organizations would not be feasible without the backup of the copier and the computer. In 1986 businesses in the United States bought 200,000 photocopiers, and this market is expected to increase for some years to come. It is difficult to comprehend that in 1986 about 45 billion pieces of bulk mail alone were handled by the U.S. Post Office. Notwithstanding the popularity of electronic mail, facsimile machines are materializing by the millions and spewing forth even more paper.

One factor that further encourages the storage of data on paper is that it is unsafe to assume that electronically stored records will be readable for even a small fraction of the 200- or 300-year lifetime of acid-free paper (National Research Council, 1986). Even if the data are imprinted on poor paper, it is always possible to photocopy it and obtain a better copy than the original, before the sulfite sheet crumbles into its acid grave. Evidence of our insecurity about electronic memory is that, although 90 percent of securities trades now take place through electronic means, they are, as one can surmise, backed up by mountains of paper.

So perhaps it is not surprising that in the information era, the trees of the world are at risk. Moreover, the equivalent of about 1,500 pounds of petroleum is required to make a ton of paper (Tenner, 1988). One wonders which will last longer—energy or the trees. Imagine the implications for the environment if a cost-effective, but nonbiodegradable, plastic substitute were found for paper! Parenthetically, we might add that biotechnology, operating at the genetic level, might be expected to bring about dematerialization to an extent even beyond that anticipated for the information technologies. However, if the end result is not only a new gene but also an enormous "supercow," then the effect again may well be materialization.

The increase in paper waste is related closely to the broad arena of efficiency in use as well as recycling. Examination of municipal and industrial waste (solid and liquid) shows that the annual generation rate per capita in the United States was estimated in the mid-1970s at approximately 3,600 pounds (Tchobanoglous et al., 1977). Japan was closest to the United States with an estimated average of 800, followed by the Netherlands at 680, and the Federal Republic of Germany at 500. The reliability and comparability of the estimates are uncertain because Cointreau (1982), for example, shows only a factor of two difference in daily per capita waste generation between New York and cities such as Hamburg and Hong Kong. Moreover, comparable estimates of which we are aware do not include emissions of environmentally important substances such as gaseous air pollutants or carbon dioxide. The human race now discharges to the atmosphere more than 5 billion tons of carbon dioxide annually, or 1 ton per person.

The considerably smaller rates of waste generation in other industrial countries are often attributed to either a lower consumption rate of goods or a more serious effort to recover and reuse the wastes (Tchobanoglous et al., 1977). In this connection, it would be instructive to examine questions such as how much paper is sold per capita in the United States, what fraction of a newspaper is recovered, whether more envelopes can be designed for reuse, what fraction of paper wasted is still usable, and what fraction of paper available for recycling is actually recycled. According to one estimate (Hagerty et al., 1973), only 24 percent of the 47 million short tons of recyclable paper in U.S. solid waste was recovered in the early 1970s.

Although paper makes up the greatest fraction of solid waste (30–35 percent), it has one of the lowest recovery rates, following textiles (17 percent) and zinc (14 percent). These low recovery rates are more than likely due to economic reasons. A recent *Wall Street Journal* article (Paul, 1989) stated, "The bottom, has fallen out of the market for recycled newspapers, exacerbating the nation's already critical garbage problems." It is reported that just a few months ago municipalities were receiving as much as $25 per ton for their newspaper waste, whereas they must now pay about $5 to $25 per ton to have old newspapers hauled away. This situation is counter to the myth that recycling should always make money. In this volume, Ayres, Ausubel, and Lee each argue that perceived scarcity of physical resources usually leads to technological substitutions. If substitution is not possible, then recycling is considered. From a purely

TABLE 6 Scrap Use in the United States

| Material | Total Consumption (million short tons) | | | Percentage of Total Consumption in Recycled Scrap | | |
|---|---|---|---|---|---|---|
| | 1977 | 1982 | 1987 | 1977 | 1982 | 1987 |
| Aluminum | 6.49 | 5.94 | 6.90 | 24.1 | 33.3 | 29.6 |
| Copper | 2.95 | 2.64 | 3.15 | 39.2 | 48.0 | 39.9 |
| Lead | 1.58 | 1.22 | 1.27 | 44.4 | 47.0 | 54.6 |
| Nickel | 0.75 | 0.89 | 1.42 | 55.9 | 45.4 | 45.4 |
| Steel/iron | 142.40 | 84.00 | 99.50 | 29.4 | 33.4 | 46.5 |
| Zinc | 1.10 | 0.78 | 1.05 | 20.9 | 24.1 | 17.7 |
| Paper | 60.00 | 61.00 | 76.20 | 24.3 | 24.5 | 25.8 |

SOURCE: Institute of Scrap Recycling Industries (1988).

economic standpoint, high-grade resources are exploited before lower grade resources and recycling are considered economically viable.

An overall view of scrap usage in the United States during 1977–1987 is shown in Table 6, where data on total consumption and percentage of total consumption in recycled material for a number of metals, as well as paper, are presented. Among the metals, there was an increase in total consumption of aluminum, lead, and nickel over the 10 years examined, whereas there was a decrease in steel and iron, as mentioned earlier. During this same period the percentage of total consumption in scrap increased for aluminum, lead, and steel. Zinc and paper have the lowest percentage of total consumption in recycled scrap, namely, 17.7 and 25.8 percent, respectively, in 1987. It is difficult to see exactly what correlations may exist or the underlying reasons for the observed variations. The availability of scrap might be expected to depend on total consumption, but it is also a function of usage, costs, and other factors.

Another question to be raised in connection with the economics of consumption and disposal is what the "true" cost of consumption and processing of the generated waste is to society. What is the true cost of burning fossil fuel for transportation when, for example, the finiteness of resources and consequent long-term damage to the environment are considered? Should high-grade resources be made available at much higher cost so that profits may be reinvested toward development of the capital and the knowledge to permit the use of lower-grade resources and the development of technological substitutes? What is the actual disposal cost of municipal and industrial wastes? To what extent is the cost of waste collection subsidized by different societies and different segments of a society?

Would a higher cost for garbage collection effectively encourage recycling, sorting recyclable materials at the generation source, and dematerialization? Would it encourage more illegal dumping? Can society truly afford to continue functioning in its present "throwaway" mode of products such as food, clothing, diapers, and shoes, as well as watches, radios, flashlights, light bulbs, cameras, calculators, pens and pencils, razors, knives, spoons, and forks?

A practice potentially very risky to society is the emission of chlorofluorocarbons (CFCs) to the atmosphere (see Glas and Friedlander, this volume). Projected depletion of the ozone layer, attributed to the environmental release of CFCs, resulted in the U.S. ban of nonessential CFC aerosol propellants in the mid-1970s. Combined release of CFC-11 and CFC-12 in the United States traditionally accounted for about one-third of the total worldwide release of these substances. The aerosol ban, however, resulted in only a gradual and temporary decline of production and emissions levels (see Glas, this volume, Figure 5), because CFC-11 and CFC-12 have had other, growing, nonaerosol industrial applications such as in refrigeration, air-conditioning, cleaning electronic and computer equipment, and foam manufacturing (Warhit, 1980). According to a 1987 international treaty the industrial countries agreed to cut CFC production in half by the year 2000. In March 1989 there was a conference in London, attended by over 100 nations, at which a proposal for the total elimination of CFCs by the year 2000 was entertained. Although it certainly seems prudent to reduce or eliminate CFC use, one wonders whether their elimination may yet result in further materialization, for example, through a need to have bulkier refrigerators again.

Lead is another example of a substance whose wide use presents a cleanup problem. Lead-containing aerosols, paint, and vehicular exhaust are among major sources of lead in the environment. It has been estimated that an effective program to reduce exposure to lead paint from the interiors of the nation's housing stock would cost between $28 billion and $35 billion (Chapman and Kowalski, 1979). Although ingestion of lead-based paint chips is regarded as the major cause of lead poisoning in children, lead exposure results from a combination of sources, including automotive lead emissions. It is estimated that 70 percent of the lead in gasoline is emitted into the atmosphere and that this accounts for about 90 percent of airborne lead emissions (Boggess and Wixson, 1977).

In the 1970s the U.S. Environmental Protection Agency (EPA) enacted a phased reduction schedule for the lead content of gasoline that has resulted in installation of lead-intolerant catalytic converters in virtually all cars produced in the United States. The national average lead content of all grades of gasoline declined from about 2.5 grams per gallon in 1968 to less than 0.1 gram per gallon in 1988, and sales of unleaded gasoline have increased consistently. Lead was introduced as an antiknock additive to gasoline in the 1930s to increase the efficiency of automobile engines. As such, lead may have contributed to the dematerialization of cars in terms of either weight or energy. But we did not foresee sufficiently that the increasing quantity of lead in our environment would itself become a serious problem.

In a recent study the EPA (1988) identified some 30 broad categories of environmental problems (see Frosch et al., this volume, Table 1) and ranked the seriousness of these problems according to the risk they posed to the population in terms of total incidence of disease and other factors. The risks considered included cancer risk, noncancer health risks, ecological effects, and welfare effects such as materials damage to industrial, agricultural, commercial, and residential properties, among others. Lead and CFCs along with, for example, sulfur dioxide, suspended particulates, carbon monoxide, and nitrogen oxides were included in three air pollutant categories regarded as having relatively high risks. Industrial dematerialization would have a significant impact on reduction of the various risks associated with these air pollutants.

Other problems evaluated in the EPA report in which materialization is a central factor include nonhazardous municipal and industrial waste, as well as mine waste. Discharges of direct and indirect effluents and municipal sludge into surface waters and wetlands are also among the high-risk problems that might be associated with materialization. In this connection, discharges of sludge and medical waste into the oceans are pressing problems with high news visibility.

During the past few years, the Atlantic Ocean has been regurgitating progressively more garbage and waste onto the beaches of the northeastern United States, especially around New York and New Jersey. Included in the dumping that causes this shocking situation are some 500,000 pounds of medical waste per week from New York City alone. Examples of materialization resulting from medical technology are the plastic throwaway hypodermic syringe and throwaway needles. In the old days, glass syringes and high-quality surgical steel needles were sterilized and used many times over. At present, syringes, for many good reasons, are used once and thrown away, as is much other medical material.

With the burgeoning of hazardous medical waste, the disposal task, especially at hospitals, becomes complex and expensive. This unquestionably leads to illegal dumping to cut costs and avoid demanding procedures. It is difficult to believe that clinic and hospital authorities are not aware of the dangers associated with illegal disposal. The midnight dumping of medical wastes raises the question of the role of the entire spectrum of "criminal" activity in our society with regard to transport and disposal of materials. Attempts are being made to determine at what point in the disposal chain the system breaks down. The solution to this type of complex problem must, of necessity, have an ethical component, with better values placed over and above such considerations as cost-effectiveness.

Although no recycling process is 100 percent efficient, recycling is a promising means of dematerialization. The construction industry is one of the major generators of solid waste. What fraction of construction waste is reusable? To what extent are brick, wood, steel, and asphalt reused? In general, a more thorough examination of practices in the construction industry regarding waste generation and processing is warranted in studies of dematerialization. How much waste is generated in construction activities such as paving roads and building houses? What happens to the waste from building construction or from demolished buildings? What determines whether a building should be demolished or renovated? What fraction of buildings is demolished as a result of safety considerations or to be replaced by a larger structure for economic reasons? What is the potential for recycling materials resulting from demolition operations, as well as various construction activities? To what extent do construction and demolition "activate" environmentally significant materials? The embalming of no-longer-usable nuclear power plants is an interesting case of permanent structural materialization.

In a recent essay, Marland and Weinberg (1988) make a powerful case for a life-cycle approach to infrastructure systems, exploring connections between quality of service provided and aging of facilities. They ask three fundamental questions about a variety of infrastructure systems: What actually is the characteristic longevity of a given infrastructure? How long could it last? How long should it last? This first attempt at a demography of infrastructure needs to be pursued in many areas in connection with materialization. From an environmental perspective, what could and should be the design life of everything we create? In the area of nuclear materials we are accustomed to asking long-range questions about how materials will be transported, stored, and disposed of. Such a life-cycle perspective might be applied usefully to other materials as we contemplate transforming them for human purposes, and thus provide guidance about instances in which dematerialization rather than materialization should be the eventual objective. More generally, it might be useful to undertake materialization impact assessments for selected new products and activities. Furthermore, the interplay between dematerialization and transportation costs in terms of weight and bulk should be examined.

The questions raised and discussions set forth in this chapter point to a number of overall objectives, namely, to single out the important driving forces behind trends in materialization and dematerialization, to determine whether on a collective basis such forces drive society toward materialization or dematerialization, and to assess the environmental implications of these long-term trends. Many questions remain to be answered quantitatively; for example, how much basic material and how many of each major product are used per capita over time and what is the lifetime of various manufactured products? If we consider that for every person in the United States we mobilize 10 tons of materials and create a few tons of waste per year, it is clearly important to gain a better understanding of the potential forces for dematerialization. Such understanding is essential for devising strategies to maintain and enhance environmental quality, especially in a nation and a world where population and the desire for economic growth are ever increasing.

## ACKNOWLEDGMENT

The authors gratefully acknowledge assistance and comments from Walter Albers, Robert Ayres, Gerald Culkin, Denos Gazis, Shekhar Govind, Ruth Reck, Richard Rothery, and Hedy Sladovich.

## NOTES

1. In an essay published in the proceedings of the Sixth Convocation of the Council of Academies of Engineering and Technological Sciences, Colombo (1988, pp. 26–27) makes the following observation:

    [E]ach successive increment in per capita income is linked to an ever-smaller rise in quantities of raw materials and energy used. According to estimates by the International Monetary Fund, the amount of industrial raw materials needed for one unit of industrial production is now no more than two-fifths of what it was in 1900, and this decline is accelerating. Thus, Japan, for example, in 1984 consumed only 60 percent of the raw materials required for the same volume of industrial output in 1973.

    The reason for this phenomenon is basically twofold. Increases in consumption tend to be concentrated on goods that have a high degree of value added, goods that contain a great deal of technology and design rather than raw materials, and nonmaterial goods such as tourism, leisure activities, and financial services. In addition, today's technology is developing products whose performance in fulfilling desired functions is reaching unprecedented levels. . . . One kilogram of uranium can produce the same amount of energy as 13 U.S. tons of oil or 19 U.S. tons of coal, and in telecommunications 1 ton of copper wire can now be replaced by a mere 25 or so kilograms of fiberglass cable, which can be produced with only 5 percent of the energy needed to produce the copper wire it replaces.

2. It would be interesting to venture calculations about the significance for materialization of the increasing average height and weight of humans, even though this effect is small compared with that of present population growth. The increase directly expands needs for textiles and food, as well as creating pressure for larger vehicles and dwellings.

## REFERENCES

Boggess, W. R., and B. G. Wixson. 1977. Lead in the Environment. Report NSF/RA-770214. Washington, D.C.: National Science Foundation.

Chapman, E. R., and J. G. Kowalski. 1979. Lead Paint Abatement Costs: Some Technical and Theoretical Considerations. Washington, D.C.: U.S. Department of Commerce.

Cointreau, S. J. 1982. Environmental Management of Urban Solid Wastes in Developing Countries. Washington, D.C.: World Bank.

Colombo, U. 1988. The technology revolution and the restructuring of the global economy. Pp. 23–31 in Globalization of Technology: International Perspectives, J. H. Muroyama and H. G. Stever, eds. Washington, D.C.: National Academy Press.

Evans, L. 1985. Car size and safety: Results from analyzing U.S. accident data. Pp. 548–555 in Proceedings of the Tenth International Conference on Experimental Safety Vehicles, Oxford, U.K., July 1–5, 1985. Washington, D.C.: U.S. Government Printing Office.

Hagerty, D. J., J. L. Pavoni, and J. E. Heer. 1973. Solid Waste Management. Environmental Engineering Series. New York: Van Nostrand Reinhold.

Hibbard, W. R. 1986. Metals demand in the United States: An overview. Materials and Society 10(3):251–258.

Institute of Scrap Recycling Industries (ISRI). 1988. Facts—1987 Yearbook. Washington, D.C.

International Energy Agency (IEA). 1987. Energy Conservation in IEA Countries. Paris: Organization for Economic Cooperation and Development and IEA.

Marland, G., and A. M. Weinberg. 1988. Longevity of infrastructure. Pp. 312–332 in Cities and Their Vital Systems, J. H. Ausubel and R. Herman, eds. Washington, D.C.: National Academy Press.

Miller, E., ed. 1988. Ward's Auto World. Detroit, Mich.: Wards Communications.

Motor Vehicle Manufacturers Association. 1982. Information on the Use of Various Materials in the Automotive Industry. Detroit, Mich.: Policy Analysis Department.

National Academy of Engineering. 1985. The Competitive Status of the U.S. Steel Industry. Steel Panel Committee on Technology and International Economic and Trade Issues. Washington, D.C.: National Academy Press.

National Research Council. 1986. Preservation of Historical Records. Commission on Engineering and Technical Systems. Washington, D.C.: National Academy Press.

Paul, B. January 25, 1989. Market for recycled newspapers in U.S. collapses, adding to solid waste woes. Wall Street Journal B4(E).

Stark, H. A., ed. 1988. Ward's Automotive Yearbook. Detroit, Mich.: Wards Communications.

Tchobanoglous, G., G. H. Theisen, and R. E. Eliassen. 1977. Solid Wastes—Engineering Principles and Management Issues. New York: McGraw-Hill.

Tenner, E. March 9, 1988. The paradoxical proliferation of paper. Princeton Alumni Weekly.

U.S. Bureau of the Census. 1975–1985. Statistical Abstract of the United States. Washington, D.C.: U.S. Government Printing Office.

U.S. Environmental Protection Agency. 1988. Unfinished Business: A Comparative Assessment of Environmental Problems. Springfield, Va.: National Technical Information Service.

Warhit, E. 1980. Regulating chlorofluorocarbon emissions: Effects on chemical production. Report EPA-560/12-80-0016. Washington, D.C.: U.S. Environmental Protection Agency.

Westerman, R. R. 1978. Tires: Decreasing solid wastes and manufacturing throughput. Report EPA-600/5-78-009. Cincinnati, Ohio: U.S. Environmental Protection Agency.

Williams, R. H., E. D. Larson, and M. H. Ross. 1987. Materials, affluence, and industrial use. Annual Review of Energy 12:99–144.

# Regularities in Technological Development: An Environmental View

JESSE H. AUSUBEL

*Forward, forward let us range;
Let the great world spin forever down
the ringing grooves of change.*

*Tennyson, "Locksley Hall," 1842*

Accept for the moment that there are long-term regularities in technological development. Suppose that the evolution and use of both individual technologies and entire technological systems are sometimes tightly consistent and predictable over decades and generations. Then, we can know with confidence some important sources of future stress on the environment and, equally, what technologically based stresses may fade, largely through natural advancement of the industrial economy. The thesis of this chapter is that, in fact, there are such long-term regularities in technological development and that these deserve more attention for the important implications they have for environmental concerns.

Let me draw you back a century to a forgotten episode of environmental history. The photographs in Figure 1 show the key material in terms of bulk in the massive expansion of the railroads in the nineteenth century. It is not widely remembered that railroads, usually associated in our minds with coal and iron, were largely wooden systems in their early development. The "iron horse" was something of a misnomer. Fuel for locomotives was

FIGURE 1 *Above*: track being laid in 1887 across the vast Montana Territory; *below*: close-up of track laying. Records of 8 miles per day were set by using huge volumes of wood, large labor forces, horses, and wagons. SOURCE: Burlington Northern Railroad Company.

wood, cars were wood, some of the rails were wood, trestles were wood and most important, crossties were wood. About the turn of the century President Theodore Roosevelt spoke as follows:

> Unless the vast forests of the United States can be made ready to meet the vast demands which this [economic] growth will inevitably bring, commercial disaster, that means disaster to the whole country, is inevitable. The railroads must have ties. . . . If the present rate of forest destruction is allowed to continue, with nothing to offset it, a timber famine in the future is inevitable.
>
> Speech to the American Forest Congress, 1905 (quoted in Olson, 1971, p. 1)

An industry leader in 1906 described the railroads as the "insatiable juggernaut of the vegetable world" (Olson, 1971). Such images were echoed in Argentina, India, the Middle East, and parts of Europe as railway networks were extended at the expense of local forests. In the United States, prevention of destruction of forests was proposed through a range of both supply and demand strategies. It was proposed to cover Kansas with a catalpa forest dedicated to supplying crossties. Railroad companies were asked to plant trees along the rail right-of-way to have a renewable stock of timber for ties. Better management of remaining forests was seen as urgent; in fact, the Forest Service was in large part built under Gifford Pinchot in this era in response to the railroad-induced crisis.

What eventually contributed most to averting the forecast crisis were, initially, creosote and other technologies for preserving crossties and, later, especially in Europe, replacement of wood by concrete ties. As is evident from Figure 2, around the time of the peak of the perceived crisis, a technological solution was already penetrating the market for crossties. Preservation technologies tripled the life of ties, and within a couple of decades, the juggernaut of the vegetable world was satiated. In fact, in the 1920s the railroad network itself reached saturation (see Figure 5), so that demand for both new and replacement ties decreased. Railroads today are almost always described as environmentally benign. So, in the railroad timber story, new technologies are both cause and cure of environmental problems. The new transportation system placed intense demand on natural resources, and innovations in turn alleviated the demand to the extent that today the issue is obscure or forgotten.

At this point it is necessary to make a brief methodological comment. A premise of this chapter is that, as suggested by Figures 3 and 4, sociotechnical systems, like biological systems, often grow according to basic patterns well-described by S-shaped curves, in particular, logistic functions (Hamblin et al., 1973; Lotka, 1956; Montroll and Goel, 1971; Volterra, 1927). In the simplest case, technologies, like biological organisms in constrained environments, proceed through a life cycle of early development through rapid growth and expansion to saturation or senescence. Often two technologies are in competition for an "econiche," that is, the market; then a logistic

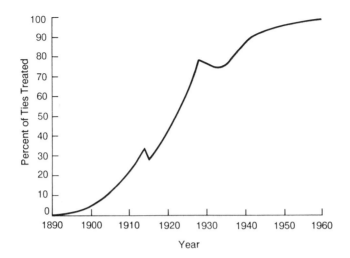

FIGURE 2 Percentage of crossties manufactured in the United States treated with chemical preservatives. It is interesting to note that the innovation penetrated the market in a characteristic S-shaped curve, disturbed only temporarily by world war and depression. SOURCE: After Olson (1971).

substitution model applies where a new technology replaces the old and the status of the system is described by the changing fraction or share of the market held by the technologies (Fisher and Pry, 1971). When more than two technologies are competing for a market, a generalized version of the logistic substitution model can be used (Marchetti and Nakicenovic, 1979; Nakicenovic, 1988, pp. 212–220).[1] Logistic functions and logistic substitution models are a compact way of presenting data on the history of technology and are used frequently in the following sections of this chapter. However, numerous methods exist to explore quantitatively the existence of patterns in sociotechnical phenomena (Montroll and Badger, 1974), and the method used most frequently here should be taken simply as indicative of the value of extending the search for regularities by using a variety of methods.

Some examples make the case for long-term regularities and also point out hazards in identifying them. Figure 5 shows the remarkably stable and parallel growth of three major systems of transport infrastructure in the United States: canals, railroads, and paved roads. For each of these transport infrastructures it would apparently have been possible relatively early in the life history of the system to make quite an accurate prediction about its eventual size and scope. Such vision in turn may be translated into conjectures about environmental problems and technological opportunities, indeed about technological necessity. For example, it could have been

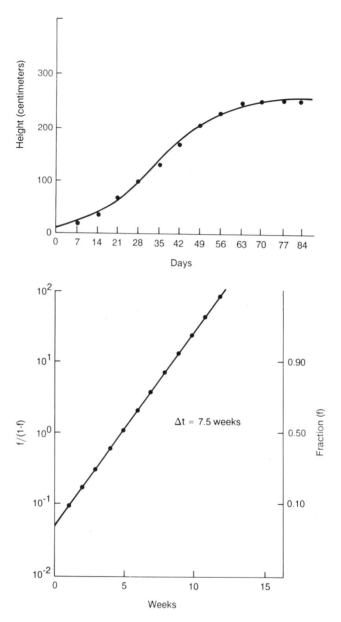

FIGURE 3 The upper panel shows the growth of a sunflower, measured in height, precisely charting a logistic curve (Reed and Holland, 1919). The lower panel shows the same data in linear transform, which is sometimes easier to employ for visual inspection and emphasizes the predictability of the process once established. For example, the ultimate height of about 260 centimeters could be estimated quickly with the linear transform. The "$\Delta t$" refers to the time for the process to go from 10 to 90 percent completion, in this case 7.5 weeks (see Lotka, 1956; Marchetti, 1983).

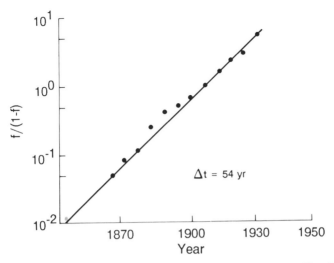

FIGURE 4 Growth of the length of wire for the U.S. telegraph system. Notwithstanding the battles involving Western Union and its predecessors and competitors, and all the associated economic and regulatory issues, the telegraph system spread its branches just as a sunflower plant grows. It is also interesting to note that the time the system required to reach its full extent ($\Delta t$) was slightly more than 50 years. SOURCE: Marchetti (1988).

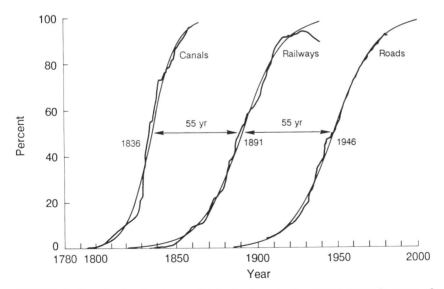

FIGURE 5 Growth of major transport infrastructures in the United States in terms of percentage of length of final saturation level. Both actual data and best-fit logistic curve are shown. The midpoint of the growth process is also shown. SOURCE: Grübler (1988).

TABLE 1  Vehicular Pollution

| Means of Transport | Pollutant | Emissions (grams per mile) |
|---|---|---|
| Horses | Waste, solid[a] | 640 |
| | Waste, liquid[b] | 300 |
| Automobiles[c] | | |
| | Hydrocarbons | 0.25 |
| | CO | 4.7 |
| | $NO_x$ | 0.4 |

[a] Calculation based on an average production of 16 kg of solid waste per day and a range of 25 miles per day.
[b] Calculation based on an average production of 7.5 kg of liquid waste per day and a range of 25 miles per day.
[c] 1980 U.S. piston engine standards.

clear early on that a rail system of predictable dimensions would be unsustainable as a predominantly wooden technology and required innovations in materials and other areas to reach forecast dimensions. Agendas for research and for entrepreneurship might have stemmed from this analysis.

A similar argument can be made about the system of paved roads. This system was initially designed for horses and horse-drawn vehicles, preceding the widespread use of the automobile. From an environmental perspective, a road system of the dimensions that began to be built could have been catastrophic if the traffic were horses. Table 1, based on calculations made by Montroll and Badger (1974), shows that, from an environmental point of view, cars were a marvelous technological innovation, at least when they were not much more numerous than horses.

Figure 6, showing the substitution of cars for horses, emphasizes the continuity of the demand for personal transportation service and the fact that technologies or modes compete to meet such demand. When considering the intensity of problems of urban air pollution in places such as Denver, Los Angeles, and Mexico City, the time may be at hand when an improvement almost as radical as that of substituting cars for horses is needed to accommodate growth in transportation demand. It is sometimes suggested that methanol fuel or electric cars will do the trick, but methanol has few obvious advantages over gasoline used in conjunction with a catalytic converter, and a versatile and wide-ranging electric car may not be available for decades. Methane and hydrogen cars are already technologically feasible and could meet stringent new environmental constraints, but they demand

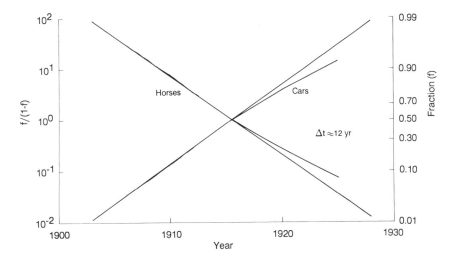

FIGURE 6 Replacement of horses by automobiles in the United States. Irregular lines are historical data; smooth lines are best fit and extrapolation. SOURCE: Nakicenovic (1988).

emergence of a substantial infrastructure of supporting service that so far is not evident. Will a breakthrough come and, if so, when?

The abrupt replacement of horses by cars shows one of the shortcomings of the type of framework presented here, namely, the difficulty of anticipating system bifurcations and fluctuations. Although the growth of overall demand for transportation as represented by horses or cars between 1900 and 1930 appears consistent in Figure 6, within 20–30 years a radical change occurred in the way that demand was met. Diesel technologies conquered steam with equal rapidity, and jets replaced propeller aircraft in about the same time. Could the timing of the introduction of such new technologies have been foreseen on the basis of a sound and transferable logic? How many in policy positions in government or industry would have believed that transformations of the transport system could occur so rapidly? Many might have recognized in 1900 that the horse-powered system was environmentally unsustainable and foreseen the concomitant need for technological solutions. I suspect that these solutions would more commonly have been believed to be incremental, for example, the breeding of horses that would be more powerful for their size (more fuel efficient) or somehow generate less waste.

It is also interesting to note long-term regularities within the automobile system, where technologies specifically employed for environmental improvement have followed typical patterns of substitution and diffusion. Figure 7 shows the adoption of emission-reducing technologies and then

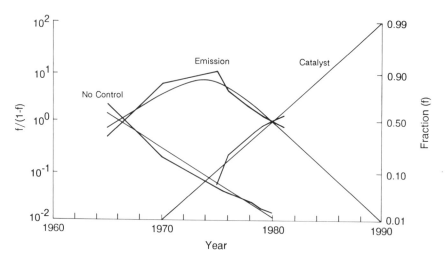

FIGURE 7 Substitution of emission controls in the U.S. vehicle fleet. The category "emission" refers to crankcase, exhaust, and fuel evaporation controls. SOURCE: Nakicenovic (1985).

catalytic converters. Identification of historically characteristic rates of such substitutions might help in setting feasible targets for future fleet improvements.

More examples of the implications of long-term regularities in technology for environment are found in examination of the transport system in its entirety (see Figure 8). If the road system is considered, it seems clear from Figure 8 (and Figure 5) that the challenge over the next many decades is maintenance and repair of a large, mature system. The road system is in fact fully grown and decreasing as a proportion of the length of the total transport system. However, we just seem to be coming to grips with environmentally sound operation and maintenance of the system that has been built. For example, with current practices and technology, the amounts of salts (close to 400 pounds per capita in the United States in 1980; Hibbard, 1986) and other chemicals that might be used for the next 50 or 100 years to keep the system ice-free are staggering. Their accumulations almost certainly pose worrisome problems for soils and water. Under the auspices of the Strategic Highway Research Program of the National Research Council (1988), technological alternatives are beginning to be explored. Accumulations of chemicals connected either with fuels that will wash off the roads or with the wearing out of tires (see Ayres, this volume) might be another issue that is now being underestimated.

Conjectures can also be offered about pressures on environment from the air transport sector. Since concerns faded in the early 1970s about

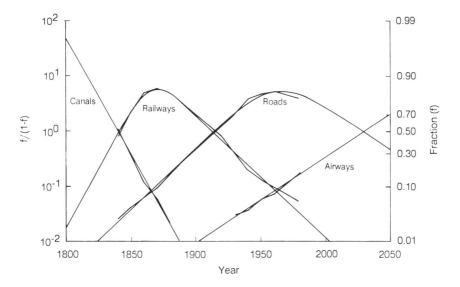

FIGURE 8 Shares of total operated intercity route mileage of competing transport infrastructures. SOURCE: Nakicenovic (1988).

stratospheric effects of fleets of supersonic transport planes (SSTs), little attention has been paid to environmental aspects of aviation. Noting the tremendous growth projected for the air transport system, one may wonder if concerns lie ahead, either in the stratosphere with a large fleet of second-generation SSTs or perhaps in a more straightforward manner in the troposphere. Could tropospheric ozone be significantly enhanced if growing emissions of nitrogen oxides ($NO_x$) by aircraft are considered? Changes might be looked for in the main travel altitude region near 10 kilometers, especially in the northern hemisphere where most air traffic occurs (Bruehl and Crutzen, 1988).

The long-term regularities identifiable in adoption of transportation technologies are paralleled in the closely related energy sector. To a considerable extent, the history of environmental and safety issues is simply the underside of the history of energy development (and agriculture). On an urban scale, 700 years of this history are recounted in *The Big Smoke* (Brimblecombe, 1987), which chronicles London air pollution since the Middle Ages and describes how improvements in technologies for burning wood and coal and for ventilation helped population density to increase and morbidity to decline.

In energy, as in transport, what is most striking is the overall consistency and stability of the evolution of the technologies favored, as illustrated in Figure 9, which shows consumption of hydrocarbon fuels—wood, coal, oil,

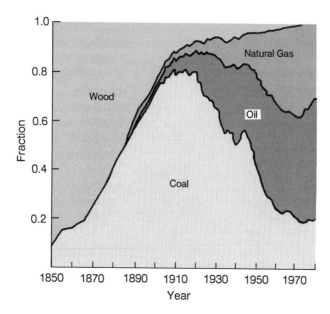

FIGURE 9 Hydrocarbon fuel consumption in the United States, fraction by fuel type, 1850–1980. SOURCE: National Research Council (1986).

and gas—for the United States in terms of market shares going back to 1850. The pattern carries with it a great deal of environmental history, for example, the deforestation that came with large-scale use of wood. Also implicit are the rise of sulfur emissions, which reached a peak in the 1920s at the apex of the coal era, and the rise of $NO_x$ emissions associated in large part with oil and the use of the automobile.

If the historical data are employed in a logistic substitution model of the kind mentioned earlier, projections of future market shares emerge (Figure 10). In this model, natural gas soon becomes the leading source of primary energy. Increasing reliance on natural gas would be a most interesting development from an environmental perspective, significantly alleviating acid rain problems. It would substantially lessen, but not eliminate, concerns about the greenhouse effect as well (Ausubel et al., 1988). Broadly speaking, we need to think about opportunities for the improvement of environmental quality offered by the possibility, indeed the likelihood, of a large role for natural gas.

At a more abstract level, one of the most interesting and best-established trends in the energy area is the substitution of hydrogen for carbon in the chemical soup that has been used to generate most energy for the past 150 years. If wood, coal, oil, and gas are all examined simply as mixtures of carbon and hydrogen atoms, then, as Figure 11 shows, global

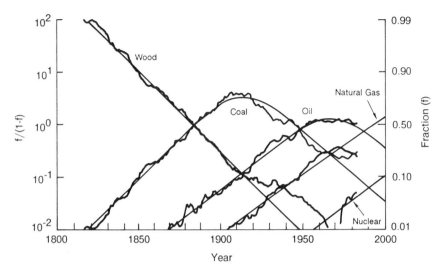

FIGURE 10 Primary energy substitution in the United States, expressed as "market share" according to the logistic substitution model, 1820–2000. The natural gas curve refers to gas resources not associated with oil exploration. SOURCE: Grübler and Nakicenovic (1988).

society has been moving steadily toward an economy running on natural gas and eventually on hydrogen. As discussed by Lee (this volume), the opportunities to evolve in the next decades beyond hydrocarbon fuels appear timely.

So far no reference has been made to "long waves," the cycles of about 50 years that seem to have characterized the world economy for the past 200 years (Freeman et al., 1982; Grübler, 1988; Nakicenovic, 1984; Schumpeter, 1939; Van Duijn, 1983). There is much disagreement about the strength of the signal that emerges in analyses of long time series of data of technological and economic phenomena that may be indicative of long waves. Figure 5 does show that a sequence of major transport infrastructures emerged at roughly 50-year intervals. Figure 9 shows that the characteristic time required for a major energy technology to capture or lose a leading role in the energy marketplace is also about 50 years. Synchronization of the diffusion of several major technologies would logically lead to periods of especially aggressive transformation of the environment and equally to "seasons of saturation" (Grübler, 1988), when environmental management might revolve more around accommodating mature systems (such as the interstate highway system).

Analysis of the evolution of energy demand shows two pulses of growth, each lasting 40 years or more (Ausubel et al., 1988; Stewart, 1988). Figure 12, which shows these pulses, may be seen as a pair of logistic curves, the second surmounting the first. From an environmental point of view,

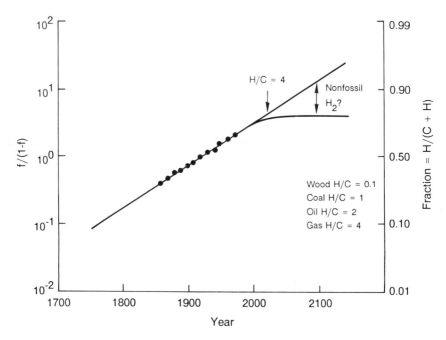

FIGURE 11  Evolution of the ratio of hydrogen (H) to carbon (C) in the world fuel mix. The figure for wood refers to dry wood suitable for energy production. If the progression is to continue beyond methane, production of large amounts of hydrogen fuel without fossil energy is required (see Marchetti, 1985).

several conjectures are worthwhile. One is that it may be possible to match each pulse with a dominant energy supply technology, coal in the first case and oil in the second. During each pulse of growth, this form of energy supply may reach environmental constraints (and other constraints as well) that limit the overall growth of the energy system. In other words, a characteristic density may be all that is achievable or socially tolerable for each form of energy within the context of a larger industrial paradigm in which that form of energy dominates. To accommodate a further increase in per capita energy consumption, a society must shift each time to a form of primary energy that is not only economically sound, but also cleaner and in some ways more efficient, especially in terms of transport and storage.

At a high hierarchical level, the cycle-adjusted view suggests that there are periods when the main orientation of the system is not so much growth as consolidation, with strong emphasis on squeezing more efficiency out of the system (a collection of technologies). At other times, the system seeks to expand rapidly and relies on introduction and diffusion of new technologies that may be "inefficient" when introduced.

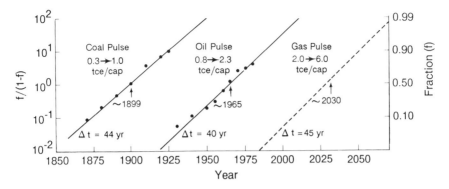

FIGURE 12 Growth pulses in world per capita energy consumption measured in tons of coal equivalent (tce). If historical discontinuities in per capita energy consumption persist, a new pulse of growth in world energy use would be expected to take off about the year 2000, which would triple per capita energy consumption from today's average world level of about 2 tce to about 6 tce (roughly half the current U.S. level). SOURCE: Ausubel et al. (1988).

It appears that we are nearing the trough of a demand cycle now. If strong demand for energy growth does not resume for another 7–10 years, as implied by the long-wave perspective, then improved energy efficiency looks like the most important near-term energy strategy, along with preparing the way for natural gas to accommodate another growth pulse (see Lee, this volume). This perspective also implies that the United States and other industrialized countries, almost all of which have sufficient capacity for electricity generation and other energy carriers in the near term, will face before the turn of the century a potential leap in energy consumption, not the steady state or low-growth world that many environmental advocates would like to see persist. To meet renewed rapid growth in demand in an environmentally sound way, gas must almost inevitably take the leading role, probably supported in particular niches by nuclear power.

It is useful to ask whether energy efficiency is always consonant with environmental improvement. At the level of particular functions such as lighting or refrigeration, it is clear that many engineering systems, indeed probably many biological systems, tend to follow steady trajectories over long periods of time toward higher efficiency (Figure 13). In most cases it may be supposed that increasing energy efficiency will also be environmentally beneficial. A counterexample is electricity. Its use is less efficient than more direct use of alternatives such as natural gas, oil, and even coal and yet is often environmentally preferred. Another counterexample, the lean-burn (Otto-cycle) engine, produces less carbon monoxide but much more $NO_x$ than a less efficient engine with a catalytic converter, which

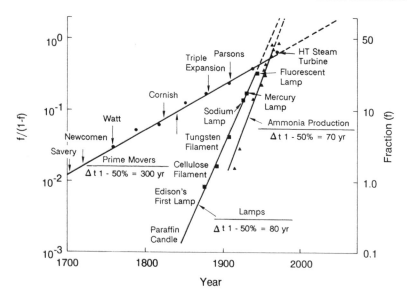

FIGURE 13 Examples of increasing energy efficiency. Prime movers, lamps, and ammonia production are measured as machines or processes for energy transformation according to the second law of thermodynamics. Original analyses are by L. M. Slesser, University of Strathclyde, Scotland. SOURCE: Marchetti (1983).

is currently more environmentally attractive at the cost of efficiency. Although the overall long-term evolution of energy systems appears to be in the direction of both efficiency and environmental compatibility, at various levels and times the system may not be optimizing for both of these objectives or they may be in conflict.

From transportation and energy, let us turn to materials, which figure prominently in the chapters by Herman et al. and by Ayres in this volume. Simple extrapolations of the kind used above have often been troublesome and unsuccessful as aids in projecting consumption of materials. As shown in Figure 14, past projections of demand for certain key materials remind us why studies such as those of Meadows et al. (1972) in the early 1970s foresaw extremely severe problems of both exhaustion of mineral resources and pollution associated with mineral use.

What happened to create the gap between the extrapolated trends and reality? Systems prove to be bounded in a variety of ways, so that exponential growth does not persist indefinitely. In the case of materials, several factors have been at work, including economic growth rates, shifts in the composition of economies from manufacturing to services, and resource-saving technologies. But most important may be smart engineering that made feasible the substitution of plastics, composites, ceramics, and optical

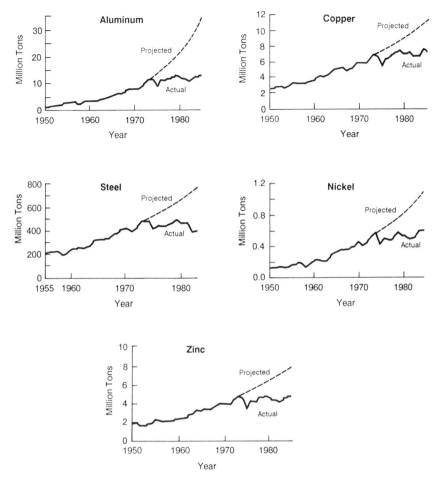

FIGURE 14 Actual materials consumption for five metals, 1950–1985, and projections made in 1970 for 1970–1985. SOURCE: Tilton (1987).

materials for metals, that is, the continuing replacement of metals with nonmetals and the associated overall decrease in metal needs.

The telecommunications sector provides a vivid example (Figure 15). In 1955, telecommunications cables were made almost entirely of copper, steel, and lead. By 1984, close to 40 percent of the materials used were plastics. If substitution of lead by polyethylene for cable sheathing had not taken place, consumption of lead by AT&T alone might have reached a billion pounds per year, an amount to create considerable anxiety from the point of view of environment, given the toxic properties of lead. Herman et al. (this volume) have examined the possible "dematerialization" of the automobile. In considerable part, the phenomenon again has to do with the substitution of plastics for metals, as implied by Figure 16.

Overall, there appears to be a decreasing dependence on common metals, perhaps combined with greater need for less common metals (Hibbard, 1986). There is also growing use of metals in the form of composites, coatings, films, and artificial structures. As Ayres (this volume) suggests, use of metals in such areas as electronics may dissipate more broadly and rapidly because many of the uses are highly dispersed and thus also entail greater complexity in recycling.

The difficulty is that good data are not readily available, and may not exist, to back up such generalizations firmly. In 1976 Goeller and Weinberg sought to develop baseline information for what they termed the "age of substitutability." One of the notions they introduced was that of "demandite," the average nonrenewable resource used by human society. They defined demandite by taking the total extraction in moles of elements such as copper and iron and selected compounds (e.g., hydrocarbons) and computing the average hypothetical chemical composition of one demandite molecule (or average mole percent composition). Goeller and Weinberg excluded renewable resources, such as agricultural products, wood, and water, from demandite but looked at them in another portion of their study.

Table 2 shows the result for the United States and for the world, for 1968, the most current year for which Goeller and Weinberg were able to perform the calculation in the mid-1970s. The dominance of hydrocarbon is striking. It is interesting that in 1968 the United States had a more favorable hydrogen-to-carbon ratio than the world as a whole, partly offsetting from an environmental perspective the fact that U.S. energy consumption is so high. Broadly speaking, the need is apparent for developing and applying concepts like "demandite" on a regular basis. With steady monitoring, such approaches might serve as indicators that would alert us to substitution processes, improving projections and reducing the likelihood of the kind of erroneous projections shown in Figure 14.

At a specific level, it is evident that, just as some environmental concerns about metals use may be decreasing, more attention must be given to plastics and paper, as also argued by Herman et al. (this volume) and Ayres (this volume). Although according to one estimate per capita use of materials in the United States remained constant between 1974 and 1985 at about 20,000 pounds per year, use of paper increased by about 25 percent to about 650 pounds, and use of plastics increased by about 40 percent to 180 pounds (Hibbard, 1986). The latter figure is an obvious and essential part of the explanation for the recent widely reported concerns about the deterioration of environmental quality at beaches in the United States and Europe (see Lynn, this volume).

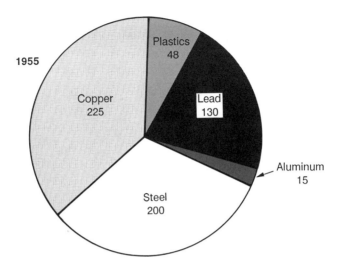
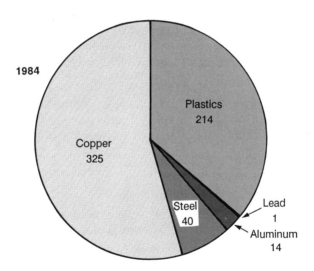

FIGURE 15 Use of materials for manufacture of telecommunications cables by AT&T Technologies, 1955 and 1984, in millions of pounds. SOURCE: Key and Schlabach (1986).

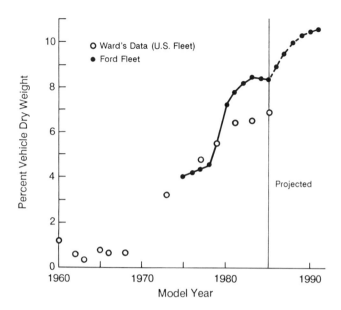

FIGURE 16 Trends in plastics content of U.S. passenger cars. Included is increased use of plastics in bumpers, fuel tanks, air cleaners, and wheel covers, but not in body panels, for which construction in plastic is also increasingly feasible. SOURCE: Gjostein (1986).

In the search for advanced materials, we may be creating materials that are virtually immortal. One wonders, for example, whether the new marvelously strong materials increasingly popular for heavy-duty envelopes are as readily recycled or biodegradable as old-fashioned paper. From an environmental point of view, electronic memory would indeed be a sought-after substitution for paper as a medium for storing information, if it could be made long-lived and reliably reproducible. Also attractive is the notion of replacing packaging itself; food irradiation, for example, may be environmentally desirable if it can significantly reduce the required volume of packaging materials.

To summarize, there is intriguing evidence of long-term regularities in the evolution, diffusion, and substitution of technologies. Understanding these regularities is of value for both environmental research and management. From numerous illustrations available in transport, energy, and materials it is evident that there is need to increase scrutiny of environmental problems and opportunities associated with growth of air transport; increasing reliance on natural gas; and disposal of plastics. Clear possibilities exist for the development of illuminating indicators, such as trends in the hydrogen-to-carbon ratio and the composition of demandite, connected to technologies and resources that would be valuable in our diagnoses

TABLE 2 Average Nonrenewable Resources Used by Man in 1968, "Demandite"

| Resource | Atomic Percent | |
|---|---|---|
| | United States | World |
| $CH_{2.14}$ | 80.22 | -- |
| $CH_{1.71}$ | -- | 66.60 |
| $SiO_2$ | 11.15 | 21.17 |
| $CaCO_3$ | 4.53 | 8.15 |
| Fe | 1.10 | 1.45 |
| N | 0.76 | 0.68 |
| O | 0.53 | 0.45 |
| Na | 0.53 | 0.45 |
| Cl | 0.53 | 0.45 |
| S | 0.23 | 0.23 |
| P | 0.08 | 0.07 |
| K | 0.07 | 0.07 |
| A | 0.11 | 0.07 |
| Cu, Zn, Pb | 0.04 | 0.04 |
| Mg | 0.04 | 0.04 |
| X | 0.08 | 0.08 |

NOTE: Here, X represents all other chemical elements: highest in order of demand are Mn, Ba, Cr, F, Ti, Ni, Ar, Sn, B, Br, Zr; others account for less than 100,000 tons per year worldwide or less than 30,000 tons per year in the United States. The term CH refers to the combination of coal, oil, and natural gas, which are all made up of carbon and hydrogen in different ratios. The subscript refers to the average hydrogen-to-carbon ratio.

SOURCE: After Goeller and Weinberg (1976).

and prognoses of environmental quality. We should not underestimate our technological ingenuity with respect to the environment nor the enormous dimensions of the systems requiring successful application of that ingenuity.

## ACKNOWLEDGMENTS

I would like to thank William Clark, Robert Frosch, Arnulf Grübler, Nebojsa Nakicenovic, and Stephen Schneider for sharing many ideas that led to this paper and Hedy Sladovich for research assistance.

## NOTE

1. Mathematically, a logistic function may be denoted by $x/(\kappa - x) = \exp(\alpha t + \beta)$, where $t$ is the independent variable usually representing some unit of time; $\alpha$ is a constant representing rate of growth; $\beta$ is a constant for the location parameter (it shifts the function in time, but does not affect the function's shape); $\kappa$ is the asymptote that bounds the function and, therefore, specifies the level at which the growth process saturates; $x$ is the actual level of growth achieved; and $(\kappa - x)$ is the amount of growth still to be achieved before the saturation level is reached. Substituting $f = x/\kappa$ in the equation expresses the growth process in terms of fractional share $f$ of the asymptotic level $\kappa$ reached; that is, the equation becomes $f/(1 - f) = \exp(\alpha t + \beta)$, the Fisher and Pry (1971) model. Taking logarithms of both sides of the equation results in the left-hand side being expressed as a linear function of time, so that, when plotted, the secular trend of a logistic growth process appears as a straight line (sometimes with perturbations). The terminology employed here is used in the figures in this chapter.

## REFERENCES

Ausubel, J. H., A. Grübler, and N. Nakicenovic. 1988. Carbon dioxide emissions in a methane economy. Climatic Change 12:245–263.

Brimblecombe, P. 1987. The Big Smoke: A History of Air Pollution in London Since Medieval Times. London and New York: Methuen.

Bruehl, C., and P. J. Crutzen. 1988. Scenarios of possible changes in atmospheric temperatures and ozone concentrations due to man's activities, estimated with a one-dimensional coupled photochemical climate model. Climate Dynamics 2(3):173–203.

Fisher, J. C., and R. H. Pry. 1971. A simple model of technological change. Technological Forecasting and Social Change 3:75–88.

Freeman, C., J. Clark, and L. Soete. 1982. Unemployment and Technical Innovation: A Study of Long Waves and Economic Development. Westport, Conn.: Greenwood.

Gjostein, N. A. 1986. Automotive materials usage trends. Materials and Society 10(3):369–404.

Goeller, H. E., and A. M. Weinberg. 1976. The Age of Substitutability or What Do We Do When the Mercury Runs Out? Report 76-1, Institute for Energy Analysis, Oak Ridge, Tenn.

Grübler, A. 1988. The Rise and Fall of Infrastructures. Dissertation, Technical University of Vienna.

Grübler, A., and N. Nakicenovic. 1988. The dynamic evolution of methane technologies. Pp. 13–14 in The Methane Age, T. H. Lee, H. R. Linden, D. A. Dreyfus, and T. Vasko, eds. Boston: Kluwer Academic Publishers.

Hamblin, R. L., R. B. Jacobsen, and J. L. Miller. 1973. A Mathematical Theory of Social Change. New York: John Wiley & Sons.

Hibbard, W. R. 1986. Metals demand in the United States: An overview. Materials and Society 10(3):251–258.

Key, P. L., and T. D. Schlabach. 1986. Metals demand in telecommunications. Materials and Society 10(3):433–451.

Lotka, A. J. 1956. Elements of Mathematical Biology. New York: Dover.

Marchetti, C. 1983. On the role of science in the postindustrial society: "Logos," the empire builder. Technological Forecasting and Social Change 24:197–206.

Marchetti, C. 1985. Nuclear plants and nuclear niches. Nuclear Science and Engineering 90:521–526.

Marchetti, C. 1988. Infrastructures for movement: Past and future. Pp. 146–174 in Cities and Their Vital Systems: Infrastructure Past, Present, and Future, J. H. Ausubel and R. Herman, eds. Washington, D.C.: National Academy Press.

Marchetti, C., and N. Nakicenovic. 1979. The Dynamics of Energy Systems and the Logistic Substitution Model. RR-79-13. Laxenburg, Austria: International Institute for Applied Systems Analysis.

Meadows, D. H., D. L. Meadows, J. Randers, and W. W. Behrens III. 1972. The Limits to Growth. New York: Universe Books.

Montroll, E. W., and W. W. Badger. 1974. Introduction to Quantitative Aspects of Social Phenomena. New York: Gordon and Breach.

Montroll, E. W., and N. S. Goel. 1971. On the Volterra and other nonlinear models of interacting populations. Reviews of Modern Physics 43(2):231.

Nakicenovic, N. 1984. Growth to Limits: Long Waves and the Dynamics of Technology. Laxenburg, Austria: International Institute for Applied Systems Analysis.

Nakicenovic, N. 1985. The automotive road to technological change: Diffusion of the automobile as a process of technological substitution. Technological Forecasting and Social Change 29:309–340.

Nakicenovic, N. 1988. Dynamics and replacement of U.S. transport infrastructures. Pp. 175–221 in Cities and Their Vital Systems: Infrastructure Past, Present, and Future. J. H. Ausubel and R. Herman, eds. Washington, D.C.: National Academy Press.

National Research Council. 1986. Acid Deposition: Long-Term Trends. Washington, D.C.: National Academy Press.

National Research Council. 1988. Annual Report of the Strategic Highway Research Program (October 31, 1988). Washington, D.C.

Olson, S. H. 1971. The Depletion Myth: A History of Railroad Use of Timber. Cambridge, Mass.: Harvard University Press.

Reed, H. S., and R. H. Holland. 1919. The growth rate of an annual plant helianthus. Proceedings of the National Academy of Sciences 5:135–144.

Schumpeter, J. A. 1939. Business Cycles: A Theoretical, Historical, and Statistical Analysis of the Capitalist Process, Vols. I and II. New York: McGraw-Hill.

Stewart, H. B. 1988. Recollecting the Future. Homewood, Ill.: Dow Jones-Irwin.

Tilton, J. E. 1987. Long-run growth in world metal demand: An interim report. Mineral Economics and Policy Program. Colorado School of Mines.

Van Duijn, J. J. 1983. The Long Wave in Economic Life. London: Allen and Unwin.

Volterra, V. 1927. Variations and fluctuations in the number of coexisting species. Pp. 65–236 in The Golden Age of Theoretical Ecology: 1923–1940, F. M. Scudo and J. R. Ziegler, eds. New York: Springer, 1978.

# 2
# The Promise of Technological Solutions

# Meeting the Near-Term Challenge for Power Plants

RICHARD E. BALZHISER

This volume frames a set of emerging multidimensional challenges to the science and technology of our energy economy and ecology. This chapter focuses on near-term challenges facing the utility industry and its engineers in accommodating to the realities of environmental, resource, and institutional constraints. It explores technological opportunities to maximize the value of past investments in meeting societal demands as well as the likelihood of finding new systems and synergies that can contribute to a healthy industry in the changing business climate in the years ahead.

The challenges the industry faces need to be understood from several different perspectives. For convenience, these challenges can be considered as a matrix in which the elements interact. The matrix can be understood first by examining each element, or cell, and then considering their interactions.

| Near term (less than 25 years) | Regional outlook (city, state, part of country) | Reasonably assured technology |
|---|---|---|
| Long term (more than 25 years) | Global outlook (hemisphere, continent, world) | Potential future options |

The first pair of cells points out the different time frames of the challenges that must be addressed. In many aspects of energy technology, the time required for extensive penetration of a market, often characterized by a logistic curve (Ausubel and Lee, this volume), is 20 years or more, because large investments in facilities and infrastructure are involved. The utility industry is learning (with urgent need) to be able to effect some changes more quickly, even when this large fleet of facilities is involved. The long term opens up many more possibilities and options in technology, but also brings many tighter constraints with respect to air, water, land use, aesthetics, and resources. The industry must, therefore, anticipate and move to cope with the long-term considerations even in what is developed and applied in the near term.

The next pair of elements in the matrix contrasts regional and global outlook. Most of what is feasible and economical in energy is determined by what is practical and convenient in a given region, whether it be a city, a state, or a section of a country. The global view has received much theoretical attention for nearly a century, but the influence of transregional and transnational constraints on local decisions and choices has begun to take effect at a practical level only in the past decade.

The third pair of matrix elements distinguishes between technology that has a reasonably firm base in knowledge and experience, and technology that may lead to potentially important applications but has either known or unexpected gaps in the knowledge and experience base.

As the research arm of the U.S. electric utility industry, the Electric Power Research Institute (EPRI) carries out development work largely in the first two cells on the top row of the matrix, that is, near term and regional. In the third column EPRI aims at evolutionary refinements in known technologies, as well as developing the knowledge base for the most promising future options. However, there is ample scope for development even of "mature" technologies that are widely used and have long been in textbooks. These developments are often in response to the challenges discussed in this volume, namely, tighter environmental constraints, or economic incentives such as an opportunity for increased efficiency or productivity in the production and use of energy and electricity. These challenges often reveal gaps in the technology base, and sometimes gaps in underlying science as well.

There are two other driving forces for development, almost at opposite ends of the spectrum of technological maturity. At one end, as much as one-third of EPRI's development is still driven by inadequacies in the fundamental knowledge base of seemingly mature technologies. This is especially evident as economic incentives for extension of useful life of major equipment intensify. Extending the knowledge base to the prediction, detection, and control or remedy for various aging and wear-out phenomena is now a major goal of research and development. (Industries concerned with aircraft, oil supply, and much of our basic infrastructure of dams, bridges, telecommunications, and buildings also see the needs, and large economic incentives, for safely extending useful lifetimes.)

At the other end of the technology maturity spectrum, EPRI supports exploratory research, not necessarily tied to any existing industrial-scale technology. This seeks to extend the knowledge base of potential future technology or methods for the production, storage, transport, or use of energy. These projects range from seeking new types of exotic composite materials, through modeling and mathematics for configuration and control of large complex computer programs, to theoretical aspects of nonneutronic fusion processes.

All of these activities include implicit judgments about the nature of the challenges for the utility industry, and for the national economy and the environmental, legislative, and administrative context in which it functions.

A key point, also made by other authors in this volume, is the need to be skeptical of forecasts. For the energy industry, skepticism means that even reasonably stable trends or widely believed forecasts need to be used in conjunction with contingency plans and options. For example, conventional wisdom is that electricity demand growth will be between 1.5 and 2.5 percent per year for the next decade at least. But 1987 and 1988 saw growth rates roughly twice this high. As another example, natural gas is generally agreed to be the most convenient fuel of choice for capacity additions over the next decade. It is clean, available, and minimizes financial risk in investment. It is convenient to install in relatively small increments for either utility or nonutility generation and for cogeneration. It continues to show attractive technical advances in lifetime and efficiency of turbines and in combined gas-steam cycle systems. Nevertheless, it would be less than prudent to neglect the possibility of price increases or supply limitations as the use of gas increases.

The enormous resource base of domestic coal can provide a vital insurance policy for energy resources supply. It is the nation's mainstay at present and is likely to continue as such, provided industry can effectively explore and demonstrate ways to make it environmentally more benign. This environmental issue for coal divides into two parts. One is to get the most useful lifetime out of the large national investment in existing coal capacity; the other is to find the next generation of technology that exploits a fundamentally different systems approach to clean coal combustion.

One development that looks especially promising involves the marriage of coal and gas technologies. This comes about from the development of efficient coal gasifiers that can feed a system using a gas turbine and a combined steam cycle. Such a system can make a technically and economically practical transition from a low-efficiency gas turbine cycle to a high-efficiency combined cycle (see Lee, this volume), and then a further transition to the still lower fuel cost by adding a coal gasifier. This is a good example of keeping robust technical options available. Such a system can stay economically useful for a wide range of possible futures on the parameters of fuel costs, capital costs, resources availability, and varying future growth rates and environmental constraints. Specifics on this line of development are discussed later in the chapter.

Other options explored at EPRI focus on improvements in the efficiency of energy use—industrial, commercial, and residential. This includes more effective use of generating capacity through regulation of load demand by on-line controls and by time-of-day pricing strategies to minimize peak loads on the system. Efficiency of use is represented, for example, by more efficient heat pumps, storage systems, and product-forming systems. Another important area is transmission. The efficient use of production capacity and energy resources demands more load-carrying capacity. With large blocks of power being transported, even small gains in the efficiency and reliability of transmission and dispatching have large economic leverage. Underground transmission and the possible use of new superconductors pose difficult scientific as well as engineering challenges. The challenges in this area are also environmental and economic. These include limitations in land use and in the availability of corridors. EPRI also does research to explore concerns over health effects of electric and magnetic fields from high-voltage transmission lines.

The combination of public concerns and regulatory constraints on investment gives further growth of nuclear generating capacity (beyond what remains in the construction pipeline) uphill prospects. However, existing capacity is a sunk cost to the economy (even if some public utility commissions disallow portions of it). It generates at about the same incremental costs as coal, and generally lower than most existing oil or gas-fired capacity. The thrust for the foreseeable future is to exploit the large experience base of operating units (nearly 100 in the United States and more than 300 worldwide) to develop a next generation of design. This promises to be economically feasible in somewhat smaller units that also have a large degree of "passive safety" and may be generally less complex to build and operate. International projects to design and develop such machines are under way, and the initial units of several versions of this family of reactors are scheduled for the early 1990s in Japan and France.

Several other chapters in this volume frame clearly the emerging challenges facing the engineering profession as more environmental limitations shift from being hypothetical to real. These challenges are heightened for the power industry by the simultaneous and sometimes conflicting pressures stemming from the elements of the matrix described earlier. They are heightened further because a society's aspirations ideally should be achieved without significantly compromising economic well-being. For the utilities, the energy/environment dilemma is especially confounding because electricity is itself such an important means to reconciliation of societal expectations, including environmental ones.

## THE ROLE OF ELECTRICITY IN ECONOMIC GROWTH AND PRODUCTIVITY

Over the centuries, auxiliary sources of energy have enabled man to leverage the economic benefits deliverable from human inputs of capital, and mental or physical labor. In the last century, electricity has amplified the economic productivity of society, initially through widespread use of motors as the muscle of our increasingly sophisticated factories. In recent years, the computer revolution has built on the versatility of electricity to move us into the electronic era, in which both man and machine work smarter and faster, achieving still higher productivity levels. This versatility reflects the "form value" of electricity—the ease of moving it and of turning it on or off (Schmidt, 1986). The form value accounts for the remarkably rapid growth of its use, despite limited thermodynamic efficiency of the Carnot cycle (cycle efficiencies of about 5 percent were common for the first decade or two) and losses in transmission.

Figure 1 shows the contrasting changes in energy and electricity intensity of our economy over the past 35 years. The use of electricity per unit of gross national product (GNP) increased until about 1975, after which it leveled off. The use of electricity has grown over the past decade at about the same rate as the economy but has been trending slightly upward again relative to GNP since 1985. Meanwhile, the intensity of energy use in our economy has declined sharply over this period since 1973. In the post-OPEC (Organization of Petroleum Exporting Countries) period, the sharp decline comes from improved use efficiencies, in the industrial and transportation sectors especially, and increasing conversion to electricity generally. The use of electricity has also reached higher efficiencies in many conventional applications. However, there are a continuing stream of new industrial applications for electricity and a continued growth in the commercial sectors that have maintained the electricity-GNP ratio, even though the ratio of energy to GNP has been falling (National Research Council, 1986).

Among the most important attributes of electricity emphasized by events of the past two decades is the ability to produce it from a wide range of resources, varying from geothermal sources and garbage on one end of the spectrum to the atom and sunlight on the other. This flexibility is central to the optimistic outlook for the increasing role of electricity in meeting our energy, environmental, and economic needs.

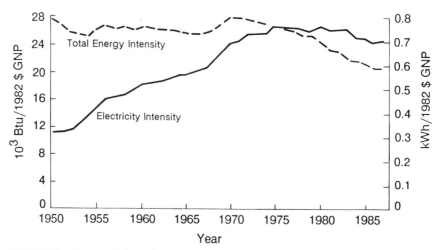

FIGURE 1 Changes in intensity of energy use in the U.S. economy, 1950–1985. Intensity is shown as the ratio of energy use, in thousands of British thermal units (Btu) and in kilowatt-hours (kWh), to U.S. gross national product (GNP) in 1982 dollars.

## IMPACTS OF REGULATIONS

The dramatic size of the political and legislative response to environmental concerns is shown in Figure 2, which chronicles laws and regulations aimed at controlling the effects of energy-related activity on air, water, and land use (Yeager and Baruch, 1987). The early challenges were the most visible and aesthetically disturbing forms of pollution, such as smoke and strip mining wastage. Pollution of the waterways, by both chemical and thermal pollutants, received attention as civic and recreational interests became affected. Finally, concerns began to focus on invisible traces of chemical constituents in gas, liquid, and solid effluents. The sensitivity of routine measurements has improved from parts per million, through parts per billion, to parts per trillion with a little more effort. These capabilities and associated concerns led to proliferation of legislation that began in the mid-1960s and continues virtually unabated today. The overall result is an enormous, uncoordinated patchwork of control requirements for smoke, air and water pollution, solid wastes, noise, and aesthetics.

The Clean Air Act (CAA), which was originally focused on human health effects much more than environmental considerations, required a seemingly simple unit operation, flue gas scrubbing, to remove sulfur dioxide ($SO_2$) from most coal plants built after 1971. These scrubbers introduced utilities to the world of chemical or process engineering and, for many companies, things have not been quite the same since (see Figure 3). Today, the nation's power plants include 62,000 megawatts of capacity with

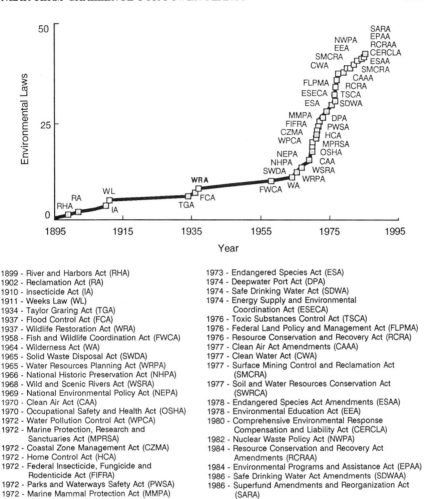

FIGURE 2  Growth in the number of U.S. environmental laws.

1899 - River and Harbors Act (RHA)
1902 - Reclamation Act (RA)
1910 - Insecticide Act (IA)
1911 - Weeks Law (WL)
1934 - Taylor Graring Act (TGA)
1937 - Flood Control Act (FCA)
1937 - Wildlife Restoration Act (WRA)
1958 - Fish and Wildlife Coordination Act (FWCA)
1964 - Wilderness Act (WA)
1965 - Solid Waste Disposal Act (SWDA)
1965 - Water Resources Planning Act (WRPA)
1966 - National Historic Preservation Act (NHPA)
1968 - Wild and Scenic Rivers Act (WSRA)
1969 - National Environmental Policy Act (NEPA)
1970 - Clean Air Act (CAA)
1970 - Occupational Safety and Health Act (OSHA)
1972 - Water Pollution Control Act (WPCA)
1972 - Marine Protection, Research and Sanctuaries Act (MPRSA)
1972 - Coastal Zone Management Act (CZMA)
1972 - Home Control Act (HCA)
1972 - Federal Insecticide, Fungicide and Rodenticide Act (FIFRA)
1972 - Parks and Waterways Safety Act (PWSA)
1972 - Marine Mammal Protection Act (MMPA)
1973 - Endangered Species Act (ESA)
1974 - Deepwater Port Act (DPA)
1974 - Safe Drinking Water Act (SDWA)
1974 - Energy Supply and Environmental Coordination Act (ESECA)
1976 - Toxic Substances Control Act (TSCA)
1976 - Federal Land Policy and Management Act (FLPMA)
1976 - Resource Conservation and Recovery Act (RCRA)
1977 - Clean Air Act Amendments (CAAA)
1977 - Clean Water Act (CWA)
1977 - Surface Mining Control and Reclamation Act (SMCRA)
1977 - Soil and Water Resources Conservation Act (SWRCA)
1978 - Endangered Species Act Amendments (ESAA)
1978 - Environmental Education Act (EEA)
1980 - Comprehensive Environmental Response Compensation and Liability Act (CERCLA)
1982 - Nuclear Waste Policy Act (NWPA)
1984 - Resource Conservation and Recovery Act Amendments (RCRAA)
1984 - Environmental Programs and Assistance Act (EPAA)
1986 - Safe Drinking Water Act Amendments (SDWAA)
1986 - Superfund Amendments and Reorganization Act (SARA)

scrubbers operating and 27,000 megawatts of scrubber capacity planned and under construction.

Congress has amended the original CAA to require a scrubber on essentially every new coal plant and is now contemplating a retrofit requirement for old coal plants to reduce acid deposition in the eastern half of the United States. Stiff emission standards are being considered for oxides of nitrogen ($NO_x$) as well, posing further challenges for coal-fired plants. Selective catalytic reduction is already commonly used on fossil stations in Japan to achieve very low $NO_x$ emissions. Its operation on U.S.

FIGURE 3 The effectiveness of flue gas desulfurization (FGD) is illustrated by the absence of deposits in outlet ductwork of coal-fired power plants. The upper panel shows solids buildup due to operating problems in the FGD absorber towers at Utah Power and Light's Hunter Station Unit 2. The lower panel shows the same duct after one year of a comprehensive FGD system optimization program.

coal plants poses much more serious engineering and operating challenges because of the lower grades of coals typically used. Combustion modification options can provide substantial $NO_x$ reductions at a fraction of the cost of catalytic reduction. The key point with respect to further environmental legislation aimed at air quality is the growing disparity between the price to the U.S. economy through higher electricity costs and the marginal air quality benefits attainable this way.

Environmental demands are not the only challenges to which utilities must adjust. Large uncertainties in long-term fuel prices and availability continue. Legislative and regulatory changes are reshaping the business of electricity supply and the obligation-to-serve concept. Meanwhile, electrical load growth continues (in some regions at phenomenal rates in 1988) and existing capacity ages. Given the growing reluctance of utilities and their regulators to add generating capacity, the need to sustain the performance of existing plants is more important than ever.

The present and planned generation mix by resource is shown in Figure 4. Coal continues to be the backbone of the current mix, representing 44 percent of the capacity and supplying 55 percent of the kilowatt-hours in 1987, much of it in the nation's heartland. A surprisingly large fraction of U.S. electricity generating capacity nationwide uses oil or gas. Because many of these units were built in days of very low oil and gas prices, much of this capacity is in peaking gas turbines that are relatively inefficient and typically lack flexibility to use alternative fuels. Today, given the higher efficiencies available with more modern combustion turbines and combined cycles, coupled with current low oil and gas prices, much of the planned generating capacity for the next decade will fall into this category. It is more difficult to anticipate and track closely, because many decisions are being put off as long as possible.

The role of independent producers that are less subject to regulation can bridge important gaps where demand growth is foreseen to outpace capacity. However, this capacity is without the obligation to serve, so a new balance may have to be struck between reliability of services and costs, as independent generation becomes a noticeable percentage of capacity in any given region.

In Figure 4, the hydroelectric component remains relatively stable, and nuclear capacity growth ends with completion of the few remaining plants in the pipeline. In the current public and regulatory climate, no chief executive officer of a public utility will even consider another nuclear plant in his remaining tenure, nor are significant additions to U.S. hydroelectric power likely. Yet, along with intensified conservation, these are among the most desirable options with respect to air quality or global warming concerns. Figure 5 shows the requirements for new power generating capacity implied by today's demand as well as 1, 2, and 3 percent compounded growth until

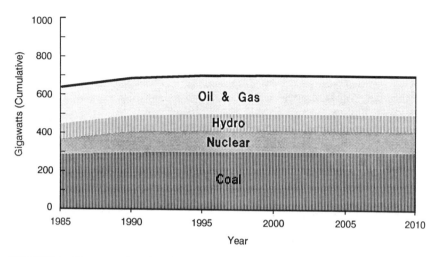

FIGURE 4 Current and projected electric power generating capacity of U.S. utilities, by fuel. Power generating construction by U.S. utilities is reaching completion with undefined commitments to meet future customer load growth. SOURCES: Utility Data Institute (1988); Federal Energy Regulatory Commission (1988).

the year 2010. Clearly, the United States is in trouble if demand parallels economic growth as it has historically. However, the real concern and more serious near-term challenge to utilities is whether they can even afford to keep some of the existing generation operating.

The shaded region in Figure 5 is a reminder that, historically, a capacity margin of at least 17 percent has been required to service users reliably at times of peak demand. Even if that performance might be improved somewhat in the future with added transmission, additional capacity will still be required.

If the aging of this capacity is considered, as shown in Figure 6, the challenges become clear. All components of the mix will be well advanced in age by the turn of the century. Like much of the aging national infrastructure, it is taken for granted as we live comfortably for the moment on the investments of the past. The operating licenses of about one-third of U.S. nuclear capacity will have expired by 2010. More than half of fossil capacity will be more than 30 years of age by 2000. Up to a third of U.S. hydroelectric capacity faces relicensing and, in some cases, serious questions of safety between now and 2010. The challenges keep coming.

The productivity and reliability of aging plants are increasingly difficult to sustain, given typical original design lifetimes of 30 years (themselves often optimistic) and the cyclic operation many units have experienced. The cost of providing for modest growth is substantial, let alone having to

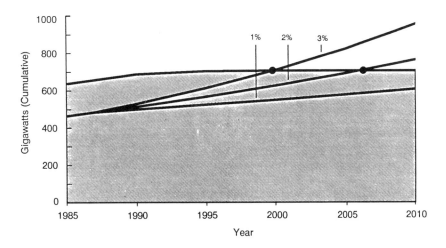

FIGURE 5  Electric power generating capacity and projected customer demand in the United States. If demand grows at 3 percent per year, it will exceed capacity before the year 2000.

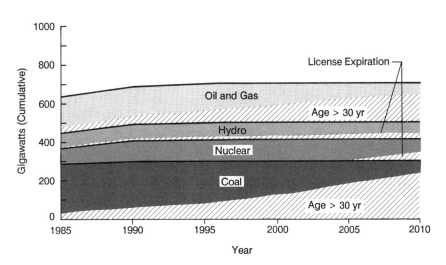

FIGURE 6  Current and projected electric power generating capacity of U.S. utilities. Cross-hatched areas show the proportion of plants more than 30 years old. Significant numbers of licenses for current nuclear power plants will begin to expire in about 15 years.

replace large portions of the existing mix. Given the fuel, environmental, and institutional uncertainties existing today, most decisions are being deferred until absolutely necessary. Short-term factors will likely dominate, which means some unattractive long-term prices may be payable in lowered reliability of service and high replacement costs.

Given the popularity of prudency reviews and major disallowances of investments after the fact by public utility commissions, there is an obvious reluctance to invest in scrubbers on aging coal plants when these could soon be made obsolete by still further requirements for the reduction of emissions. Moreover, concerns about global warming could do to coal what unlimited intervention has done to nuclear power. The investment risk of both options far exceeds that of oil and gas, and environmental intervention can drive up costs unpredictably. Commitments for new generation are logically moving to combustion turbines and combined cycles where these exposures are minimized. If U.S. gas reserves and the ability to produce and transport gas are as abundant as the utility industry projects, that is a reasonable path, provided it is possible to keep robust systems with some diversity and contingency options. Developing technology opportunities can help deal with this concern.

## TRENDS IN EFFICIENCY

An important track record of engineering achievement is shown in Figure 7, which illustrates a sustained improvement in thermal efficiency accompanied by a continuing decline in the cost of electricity over most of the last century. The reversal in both efficiency and cost coincide with imposition of more stringent emission controls for particulates and $SO_2$, as well as for thermal discharges. Energy requirements and losses associated with these control devices more than offset continued engineering improvements in both gas and steam cycle conversion technology that had pushed actual power plant thermal efficiencies above 40 percent.

The new challenge to which the industry is now responding is to reverse the negative pressures on thermal efficiency and, at the same time, improve environmental controls. Holistic approaches to past and future environmental concerns have been pursued doggedly over the past 15 years, recently with amazing success.

## CLEAN POWER FROM COAL

Two basic technologies have been adapted for producing clean power from coal: fluidized-bed combustion (FBC) and integrated gasification combined cycle (IGCC). Both technologies avoid the need for scrubbers

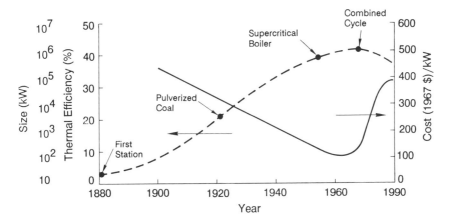

FIGURE 7  Power plant evolution. As the thermal efficiency of coal-fired steam electric generation increased from about 5 percent in the late 1800s to 35–40 percent in the late 1960s, fuel consumption per kilowatt of power produced decreased by 85 percent. During the same period, boiler size increased from 50 kilowatts to 1,200 megawatts. As a result, the cost of new generating capacity dropped from $350 per kilowatt in 1920 to $130 per kilowatt in 1967 (constant 1967 dollars), and average residential service cost dropped from $0.25 to $0.02 per kilowatt-hour. This pattern of improved efficiency and lower energy costs ended in the late 1960s, which suggests that existing power plants had approached limits set by thermodynamics, available materials, and economics. Moreover, it coincided with the increasing priority on controlling environmental pollutants.

by internalizing means for control of emissions in the combustion process. Waste products are more manageable and even salable.

Fluidized-bed combustion captures sulfur from burning coal in a bed of fluidized limestone; $NO_x$ formation is suppressed by virtue of lower temperatures in the combustor (see Figure 8). Ninety percent of $SO_2$ removal is achievable, and $NO_x$ levels are typically between 200 and 300 parts per million (ppm), or well below (about half) of today's new source performance standards.

These lower and more uniform combustion temperatures also permit increased fuel flexibility, because combustion temperatures remain below the temperature at which the ash melts and becomes slag. Adaptability to a wide range of fuels is a major advantage for FBC. It helps to ensure competitive fuel costs as prices trend upward.

These advantages are being demonstrated at a scale of approximately 80 to 160 megawatts in several key projects. Early results at Northern States Power in Minneapolis (130 megawatts), Montana-Dakota (80 megawatts), and Colorado-Ute (110 megawatts) have been extremely instructive and encouraging. Projects at TVA/Duke and American Electric Power will

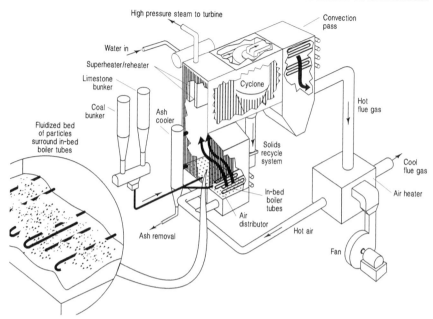

FIGURE 8 Fluidized-bed combustion (FBC) of coal. Modern FBC boilers can burn even high-sulfur coal cleanly, meeting emissions standards without scrubbers for flue gas. Forced air suspends a mixture of coal, limestone, ash, and sand, maintaining a turbulent, fluidlike bed in which some of the boiler tubes are immersed. Sulfur released by the burning coal is captured as solid calcium compounds by reaction with the limestone particles in the fluidized bed. The excellent heat transfer within the bed limits the combustion temperature, which avoids the formation of $NO_x$ and the melting of ash into slag that could block the furnace. With low-sulfur fuels, a bed of mainly coal ash or sand captures most of the residual sulfur.

explore the operational regimes of FBC pressurized to higher than atmospheric pressures.

The FBC projects have been making important but evolutionary advances in combustion technology. Meanwhile, the IGCC has made what can be considered a quantum jump forward. Successful operation of large demonstration plants at Southern California Edison's Cool Water site in California and, more recently, at Dow Chemical's Plaquemine, Louisiana, plant provide the basis for clean use of coal for the next century.

The Cool Water plant has operated with high- and low-sulfur coals over the past five years, producing a plume comparable to that of the natural gas fired combined-cycle plant adjacent to it. Emissions of $SO_2$, $NO_x$, and particulates are all below federal and state requirements as shown in Table 1 (Spencer, 1986). The vitrified solids leaving the gasifier are virtually inert, passing tough California leaching tests for contamination of drinking

TABLE 1 Cool Water Program Heat Recovery Steam Generator Stack Emissions (Utah Coal)

| Effluent | Emissions | | |
|---|---|---|---|
| | Parts per Million, volume | Pounds per Hour | Pounds per Million Btu |
| $SO_2$ | | | |
| Permit requirement | (10)[a] | 35[b] | 0.033 |
| Actual test result | 9 | 33 | 0.033 |
| $NO_x$ | | | |
| Permit requirement | (27)[a] | 140 | 0.065 |
| Actual test result | 23 | 61 | 0.061 |
| Particulates | | | |
| Permit requirement | -- | -- | -- |
| Actual test result | -- | 1 | 0.001 |

[a] Nominal translation from pounds per hour at typical conditions.
[b] Environmental Protection Agency permit requirements for Utah (SUFCO) design coal correspond to 95% sulfur removal. Permit requirement (and expected $SO_2$ emissions) for Illinois No. 6 coal test is 175 pounds per hour, corresponding to 97% sulfur removal.

water. Requirements for cooling water are less stringent in these plants than in conventional coal plants using scrubbers. Even more impressive has been the high level of availability (70 percent) realized in 1987 by this first-of-a-kind plant. If carbon dioxide ($CO_2$) must eventually be removed from plant effluents, IGCC technology can probably best accommodate this requirement—not without cost, but at costs below other coal-based alternatives.

With today's gas prices and system base load capacity, simple gas turbines for peaking are often the option of choice for most utilities. With proper planning and siting, these units can be upgraded in several stages, first from a simple cycle to a combined gas-and-steam cycle with an increase in output and in thermal efficiency. Where coal becomes the fuel of choice, an integrated gasifier can be added. This phased growth adaptability of IGCC makes it particularly attractive to utilities at the present time.

The FBC and IGCC options also have application in repowering some existing fossil units. Repowering by using either option, where feasible, achieves $SO_x$ and $NO_x$ control in addition to improving overall plant productivity—a much preferred alternative to scrubbers. The application of FBC to the Black Dog unit of Northern States Power gained 25 megawatts of capacity and 25 years of useful life.

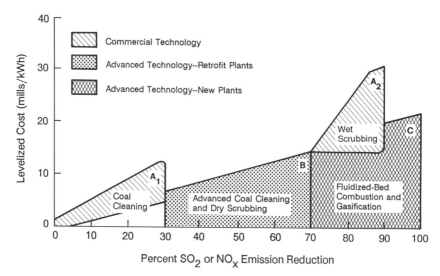

FIGURE 9 Control options for $SO_2$ and $NO_x$. Advanced coal cleaning and dry scrubbing retrofit options (B) offer cost-effective emission reductions at levels between present physical coal cleaning capabilities ($A_1$) and wet flue gas scrubbing ($A_2$) for existing commercial power plants. Improved coal use technologies both for retrofitting (B) and for new plant applications (C) can combine high levels of emission and effluent control with reduced costs and improved thermal efficiency. Levelized cost represents the present value of the life-cycle capital, operation, and maintenance costs normalized to a uniform annual cost. NOTE: 1 mill = 0.001 dollar.

Figure 9 shows the control options for $SO_2$ and $NO_x$. Options for retrofitting existing plants with advanced technology to $SO_2$ typically have lower removal potential, but also lower capital costs than wet flue gas scrubbers. In the case of $NO_x$, similar retrofit controls now under development in this country offer appreciable $NO_x$ reduction at lower capital and operating costs than current commercial scrubbers.

The essence of clean coal programs is to provide long-term holistic solutions to coal's environmental impacts, such as FBC and IGCC, and better retrofit technology. The latter is essential to mitigate legitimate environmental problems without forcing premature retirement of aging units, which are still a major part of U.S. base load capacity.

As extended lifetime performance becomes an increasingly important objective for older fossil and nuclear units, it is necessary to look for better indicators of declining safety margins or incipient failures of equipment. Ideally such deficiencies or degradation effects are detected early enough to permit the scheduling of repair work. It is also preferable to detect such conditions so as to get plants off the line in time to avoid damaging failures. One of the keys to maintaining performance is much broader

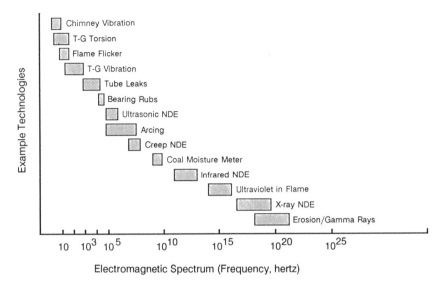

FIGURE 10 Technologies for monitoring and inspection of power generation equipment in operation. Low-frequency devices monitor or detect low-frequency phenomena, such as turbogenerator (T-G) torsion and vibration; higher frequencies are used for nondestructive evaluation (NDE), among other purposes.

coverage of equipment with on-line diagnostics, an emerging technology of great potential value to utilities. This involves both new sensor technologies and computer-assisted discrimination and diagnostic tools, including expert systems and artificial intelligence applications.

Figure 10 shows the general kinds of devices and signals that can serve to monitor equipment while it operates. Devices use a wide spectrum of frequencies to detect, for example, incipient failures in turbine rotors or hot spots in a generator or combustion turbine. Avoidance of forced outage takes on added importance as equipment is operated beyond original design lifetimes.

## CARBON DIOXIDE AND GLOBAL WARMING

Much effort has been expended over the past 15 years to deal with the mandate to make cleaner use of coal. The engineering accomplishments on this need—by many individuals and organizations—have been impressive (Conference Report, 1987; Simbeck et al., 1983). Exploiting the foregoing technology opportunities addresses all but the most recent environmental concern—global warming. If minimization of $CO_2$ emissions becomes a serious objective, the United States will most certainly want to revisit the nuclear option as well as redouble its efforts on solar energy, electric transportation, and improved efficiency in the generation and use of electricity. The United States can both engineer and afford these options. Many emerging countries can do neither and will likely continue to rely largely on coal and oil.

## CONCLUSION

Power plants are becoming more oriented to chemical processes and, therefore, more adaptable to environmental goals. Coal gasification not only produces a clean fuel for combustion turbines or fuel cells but also the building blocks for methanol and many other petrochemicals. One should recognize that coal gasification represents the core of a coal refinery with electricity as only one of many products available. Who will most successfully commercialize this technology in the future? The opportunities for adding value are large.

Given that coal can eventually be relied on for feedstock purposes, it appears that the United States can safely turn increasingly to gas for near-term energy needs. At current prices and with available combustion turbine and combined-cycle technology, it is the option of economic choice not only for peaking, but also for middle-range and some base load applications. It can be added quickly; it operates cleanly and emits less $CO_2$ per kilowatt-hour than any other fossil option. It permits time to understand the issue of global warming better without imposing costly or ineffectual $CO_2$ removal requirements. Finally, when gas prices rise, installation of coal gasifiers permits adaptation and continued productive use of the power generation investment.

The electric power industry now has a technological tool kit that can meet most of the environmental challenges identified to date. Most important, the industry and the nation have an engineering capability that will continue to adapt to changes ahead.

## REFERENCES

Balzhiser, R. E., and K. E. Yeager. 1987. Coal-fired power plants for the future. Scientific American 257(September):100–107.
Conference Report. 1987. Proceedings of Sixth Annual Conference on Coal Gasification, Electric Power Research Institute Report AP-5343-SR, October 1987. Palo Alto, Calif.
Federal Energy Regulatory Commission. 1988. Data base maintained by R. Corso, Office of Hydropower Licensing, Washington, D.C.
National Research Council, Energy Engineering Board. 1986. Electricity in Economic Growth. Washington, D.C.: National Academy Press.
Schmidt, P. 1986. Form Value of Electricity. Pp. 199–227 in Electricity Use, Productive Efficiency and Economic Growth, S. Schurr and S. Sonenblum, eds. Palo Alto, Calif.: Electric Power Research Institute.

Simbeck, D. R., R. L. Dickenson, and E. D. Oliver. 1983. Guide to coal gasification systems. Electric Power Research Institute Report AP-3109, June 1983. Palo Alto, Calif.

Spencer, D. F., S. B. Alpert, and H. H. Gilman. 1986. Cool Water: Demonstration of a clean and efficient new coal technology. Science 232(May 2):609–612.

Utility Data Institute. 1988. Edison Electric Institute power statistics data base. Washington, D.C.

Yeager, K. E., and S. B. Baruch. 1987. Environmental issues affecting coal technology: A perspective on U.S. trends. Pp. 471–502 in Annual Reviews of Energy, Vol. 12. Palo Alto, Calif.: Annual Reviews, Inc.

ically quoted from *Technology and Environment*, edited by Jesse H. Ausubel and Hedy E. Sladovich, 1989.

# Advanced Fossil Fuel Systems and Beyond

THOMAS H. LEE

The terms of the energy debate have changed dramatically over the last 15 years. Whereas the size of the fossil fuel resource base was the overriding concern of the 1970s, today the formidable challenge is how to use energy sources in ways that support social and economic development and protect the environment. To develop a strategic perspective on how to meet this challenge in the long term, it will be necessary to explore some of the misconceptions of the past that led to costly errors in energy planning. Such a review, in retrospect and prospect, will help answer the question: What happens after the fossil age?

## THE MYTH OF "RUNNING OUT OF RESOURCES"

For years, energy planners thought that the driving force for the shift from one energy source to another was resource depletion: Whatever is the most desirable and necessary resource will run out quickly or soon enough to push the movement to alternatives. This running out hypothesis has pervaded the bureaucratic, business, and scientific communities for decades. It has served as a basis for national policy, industrial policy, investment policy, and research policy. In the case of energy, it is a myth that resource depletion is the driving force for resource substitution. Studies of nonfuel minerals have led to a similar conclusion (Tilton, 1984).

For centuries, fuelwood, animal and farm waste, and animal and human muscle power were the mainstays of energy supply. Compared with contemporary energy consumption patterns, these traditional energy forms

*114*

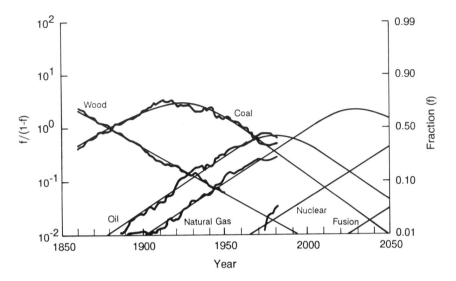

FIGURE 1 Historical and projected trends in global primary energy consumption. The amount of energy (tons of coal equivalent) from each source is plotted as a fraction $f$ of the total energy market, with $\log f/(1-f)$ as the ordinates. The smooth secular trends are the model estimates based on historical data; irregular lines are historical data. SOURCE: Marchetti and Nakicenovic (1979).

were used at low absolute levels and low densities of generation and end use. Essentially, their exploitation was not dependent on infrastructure for transformation and transport.

These patterns were altered with the emergence and intensification of the industrial revolution of the nineteenth century. As Figure 1 shows, fuelwood was replaced by coal during the latter half of the nineteenth century. Fuelwood's share declined from some 70 percent in 1860 to about 20 percent in the early 1900s at the same time that coal's share increased from 30 to almost 80 percent. Fuelwood was abandoned, not primarily because of the threat of resource depletion, but because coal mining and coal end-use technologies provided an energy source that could do what fuelwood did—and better. Although it was possible (and still is) to operate trains and ships with fuelwood and use it for shaft power and electricity, advances in coal technology made it increasingly easier, more efficient, more reliable, and cheaper to do so with coal.

However, by 1910 coal's rapid growth had ceased, with its share of the primary market peaking some 10 years later and declining in relative shares thereafter in a pattern that is almost symmetrical with that of fuelwood 50 years earlier. By the early 1960s, coal had been displaced by crude oil as the dominant fuel on the primary market, both in market shares and on an absolute basis. Today, a similar substitution pattern can be observed. Coal resources were (and still are) abundant. But with the discovery around 1860 of oil by drilling, a set of oil-related technologies began a development process that eventually led to the large-scale and efficient refining of crude oil into a broad range of products and chemical feedstock. These innovations opened the market for oil. On the end-use side, refined oil products proved to be far superior to coal for powering trains, automobiles, and aircraft; for generating electricity; and for providing residential and commercial heating. All of these end-use applications, except automobiles and aircraft, had been achieved first by use of coal. The primary radically new application opened up by the use of oil was, of course, aviation, now a large consumer of refined oil products.

Nonetheless, around 1980 crude oil peaked on the world primary market, both in terms of shares and on an absolute basis, and began to decline thereafter. As Figure 1 also shows, natural gas and nuclear energy have been steadily gaining market shares against crude oil: natural gas since the 1920s, and nuclear energy since 1970. Thus, from the historical perspective, energy substitution has been driven by the availability of a set of new technologies that enabled an alternative energy source to satisfy better the end-use demand of society.

Another point seldom mentioned is that the so-called reserves themselves are actually functions of technology. The more advanced the technology, the more reserves become known and recoverable. An excellent example is the Kern River story as described by Adelman (1987) in an address before the National Press Club.

> Kern River in California was discovered in 1899. After 43 years of production, it had "remaining reserves" of 54 million barrels. In the next 43 years of life, it produced not 54 but 730 million. At the end of that time, in 1986, it had "remaining reserves" of about 900 million barrels.

The past trend is clear: technology has been the engine of change in the energy sector. I believe that the role of technology in energy will continue to be the same as in the past, despite a shift in emphasis to environmental protection and other societal needs.

## NATURAL GAS: A BRIDGE TO THE POSTFOSSIL AGE?

More than a decade ago, the International Institute for Applied Systems Analysis (IIASA) forecast that after oil, natural gas and nuclear energy would be the dominant growth fuels over the next few decades. At that time, these predictions were a highly controversial and emotional issue. They have, however, stood the test of time reasonably well. New reserves have been discovered in many parts of the world. The consumption of methane in the world has been increasing (Figure 2), with the United

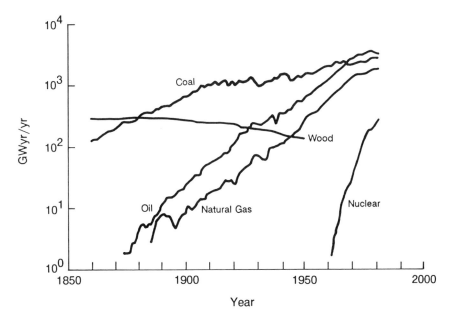

FIGURE 2  World primary energy consumption (in gigawatt-years per year). SOURCE: Grübler and Nakicenovic (1988, p. 15).

States being the single exception. The U.S. Power Industry and Industrial Fuel Use Act of 1978 (Public Law 95-620, 42 USC 8301), forbidding the use of natural gas for electricity generation, was modified in 1987. There is ample evidence to indicate that the consumption trend will reverse. A more recent gas study by IIASA indicates that after the year 2000, Europe may have to depend increasingly on natural gas for its energy needs (Rogner, 1988).

To discuss further the likelihood that natural gas will become the bridge to the postfossil age, natural gas technologies must be examined in the framework of technology life cycles. In many ways, technological systems and biological systems can be described similarly. One can define in a qualitative way three different stages in the life cycle (Figure 3): the embryonic, growth, and maturity stages. As a technology progresses through its life cycle, a number of measurable quantities evolve along S-shaped curves. The simplest S-shaped curve is described by the logistic equation. For any parameter $x$ (e.g., performance, market size),

$$dx/dt = \kappa x(y - x),$$

where $y$ is the upper limit (saturation value) for $x$, $t$ is the time, and $\kappa$ is a constant. When $x$ is plotted against $t$, it has the form of a symmetrical

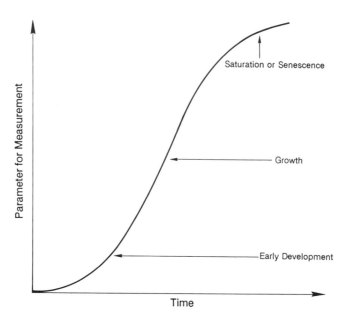

FIGURE 3  Technology life cycle.

S curve. Let $x/y$ be denoted by $f$, which is the fraction of saturation value attained by $x$. If $f/1 - f$ is plotted versus time on a semilog scale, a straight line is obtained.

Evolution of the technical performance of passenger aircraft may be used as an example (Figure 4). Instead of following closely a single straight line, there is a band, with the left line representing the performance of the best airplanes. If one had to choose between different modes of transportation for investment purposes in the 1930s, the most favorable information that was available for aviation was on the DC-3. Still, comparing the performance, cost, and personal comfort of a DC-3 with that offered by railroads, one might easily have concluded in the 1930s that the railroad would remain superior. Fifty years later, there is no convenient way to travel between coasts in the United States by rail. From looking at Figure 4, the reason is clear. The young aviation technology of the 1930s improved its performance (as measured by passenger-kilometers per hour) by more than a factor of 100 over the following 40 years. The already mature railroad technology showed no such performance improvement.

Examining energy technology in this context suggests that gas technology is still young. For years, natural gas was a by-product of oil exploration. Only recently have wells been drilled intentionally for gas exploration. By plotting logistically the drilling and production rates of so-called nonassociated gas (Figures 5 and 6), the share of gas wells is seen to increase

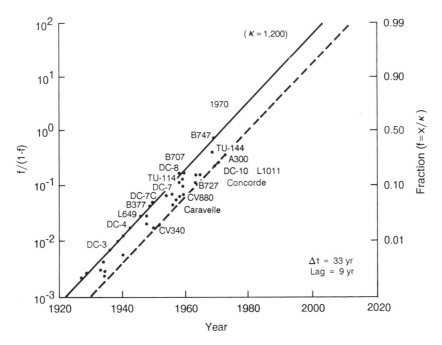

FIGURE 4 Improvement of passenger aircraft performance in thousands of passenger-kilometers per hour. Each point on the graph indicates the performance of a given aircraft when used in commercial operations for the first time. The upper curve represents a performance feasibility frontier for commercial aircraft; the performance of all other commercial aircraft at the time they were introduced was either on or below the curve. ($\kappa$ is the estimated saturation level.) SOURCE: Lee and Nakicenovic (1988).

while that of oil wells decreases. Figure 7 shows the substitution picture when nonassociated gas is separated from oil technology and shown as gas technology. If these trends continue, nonassociated gas exploration can be expected to grow for some time to come. We have also seen advances in exploration through use of remote sensing by satellites and ground truth measurements. Drilling technologies are also advancing underground, deeper and faster (Figure 8).

Perhaps the most convincing dynamic technological advance for this case is the conversion efficiency of combined-cycle systems using natural gas. In a gas turbine combined-cycle (GTCC) system, exhaust from a gas turbine is fed into a residual heat boiler that generates steam for a bottoming steam cycle. The capital cost of such a plant is considerably lower than that of a coal-fired plant or a nuclear plant (about $500 per kilowatt), and the conversion efficiency is significantly higher. It is interesting to follow the advances of that single parameter. In the 1970s, the efficiency was in the

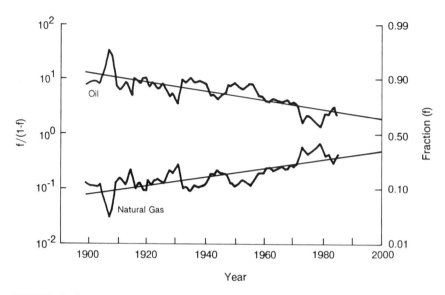

FIGURE 5 Shares of successful oil and gas wells in the United States. SOURCE: Grübler and Nakicenovic (1988, p. 27).

high 30 percent range; in the beginning of the 1980s, it went up to 45–47 percent. In 1987 when Norway was considering such a plant, the efficiency quoted by suppliers was in excess of 50 percent, an interesting example of dynamics of technology.

Despite their economic attractiveness, combined-cycle systems have not been considered a serious option in planning additions to utilities' future power generation in the United States. The reasons are many: the Fuel Use Act, lack of confidence in a reliable long-term gas supply, and lack of confidence in the performance of GTCC.

A related issue that deserves attention is the utility rate structure. Consider a hypothetical utility that has both nuclear and combined-cycle plants. If one computes the actual cost of electricity generated by the two plants, the combined-cycle plant might have an advantage. However, in daily dispatching decisions, the combined-cycle plant may not be dispatched because its fuel cost is higher than that of a nuclear plant, and the high capital cost of the nuclear plant is already in the rate base.

In dispatching, only the fuel cost counts, because the operating and maintainance costs are not a significant factor, but when the cost of electricity from a third-party generation owner is compared with that from the electric utility companies, a different standard is used. Policymakers find the total cost a better measure. Thus, a combined-cycle system that is

# ADVANCED FOSSIL FUEL SYSTEMS AND BEYOND

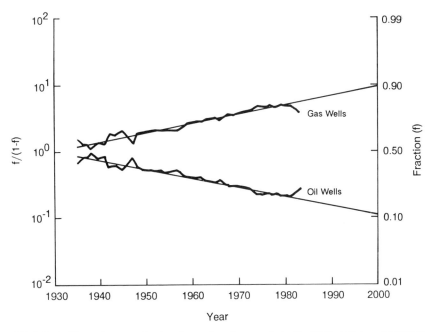

FIGURE 6  Natural gas production from oil and gas wells in the United States. SOURCE: Grübler and Nakicenovic (1988, p. 32).

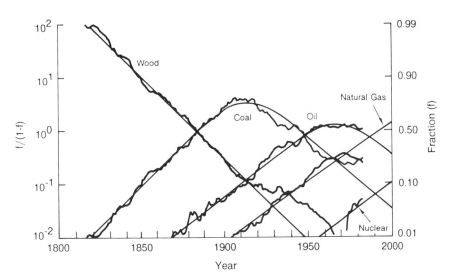

FIGURE 7  Energy substitution in the United States; gas from nonassociated wells is shown separately as gas technology. SOURCE: Grübler and Nakicenovic (1988, p. 37).

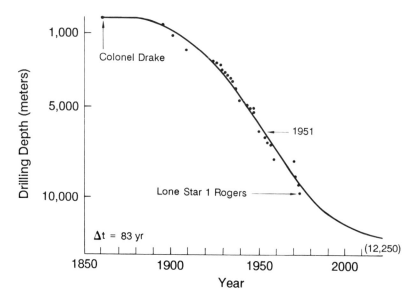

FIGURE 8 Maximum depth of exploratory drilling in the United States. SOURCE: Grübler and Nakicenovic (1988, p. 23).

not economic to dispatch in the utility system becomes a viable competitor outside the system. So, by changing one measurement arbitrarily, the deregulation of electricity generation makes sense.

The MIT Power Systems Laboratory examined the economics of combined-cycle systems from the viewpoint of total cost (Tabors and Flagg, 1986). The analytic method used was the Electric Generation Expansion Analysis System (EGEAS), developed by MIT and Stone and Webster Engineering Corporation for the Electric Power Research Institute (EPRI). The EPRI-developed Regional Utility Systems allowed extrapolation to the entire U.S. system. With a set of reasonable assumptions, the study concluded that natural gas-fired combined-cycle systems with efficiency already in hand contributed to the optimal capacity mix. In three of the six regions studied, they provided the majority or all of the optimal mix.

Concurrent with this, El-Masri (1985) made a comprehensive analysis of the efficiency of combined-cycle systems. It is important to point out that for decades, the technical development of gas turbines was influenced heavily by jet engine technology, developed by the U.S. Air Force for military purposes. Firing temperatures have increased (Figure 9), and high-temperature materials have been developed (Figure 10). But because of the weight and space limitations for aircraft applications, the exhaust temperature from the gas turbine is not optimal for the bottoming steam cycle. This situation can be improved if reheating between the gas turbine stages is considered. In 1978 a national energy savings project (the Moonlight Project) started by the Japanese government included the development of a reheat gas turbine to optimize the performance of combined-cycle systems. This occurred after Japan, in an effort to diversify energy sources, ordered several gigawatts of combined-cycle systems from the General Electric Company.

Results of the analysis done by El-Masri are shown in Figure 11, in which the combined-cycle system efficiency is plotted as a function of pressure ratios at various peak temperatures (measured by $H$, which is the ratio of the turbine inlet temperature to the ambient temperature). Each point represents the maximum efficiency configuration of a single-stage compressor and reheat turbine. The optimum number of turbine stages is indicated at each point. If one imposes a design constraint of three turbine stages (two reheats), the performances are shown by the dashed curves. With increased turbine inlet temperatures and higher compression ratios, efficiencies between 55 and 60 percent or even higher may be achieved. Thus, it is not unreasonable to say that combined-cycle systems are still in the growing phase of their life cycle.

When the possibility of high-efficiency combined-cycle systems was included in the EGEAS study, the results were indeed remarkable. The

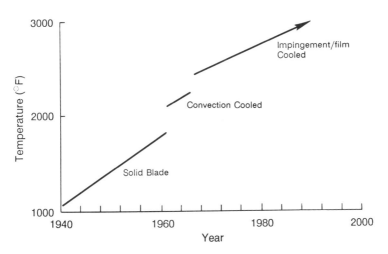

FIGURE 9  Turbine technology trends. SOURCE: Lee (1988, p. 135).

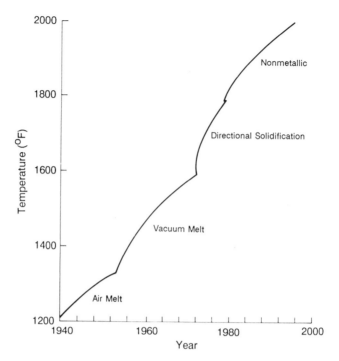

FIGURE 10 Progress in technologies for high-temperature materials development for turbine blades and disks. SOURCE: Lee (1988, p. 136).

most dramatic differences occurred in the Northeast and Southeast. Figures 12 and 13 show the expansion path for these two regions. Given the planning horizon of 15 years, only combined-cycle additions make economic sense.

This section has touched on the dynamics of only a few of the natural gas technologies. Similar attention should be given to exploration, drilling, down hole communication and control, production and transportation, and of course end use. The results of a 1986 workshop on these topics were recently published in a book entitled *The Methane Age* (Lee et al., 1988). Much remains to be done in engineering research and practice if methane is to become a bridge to the era beyond fossil fuels.

## ENVIRONMENTAL CONSIDERATIONS

We have entered an era of increasingly complex patterns of interdependence between environmental and human development. These patterns are characterized by temporal and spatial scales transcending those of most contemporary political and regulatory institutions. What were once local

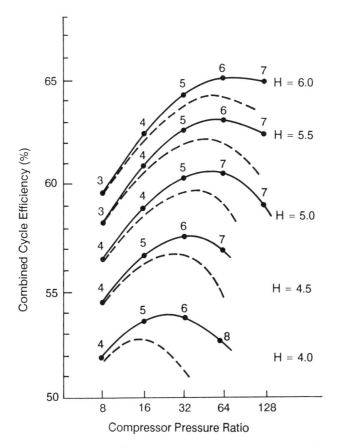

FIGURE 11 Efficiencies of combined-cycle systems at various pressures and temperatures ($H$ is the ratio of turbine inlet absolute temperature to ambient absolute temperature). Numbers above the curves indicate the number of turbine stages. Dashed line shows the performance if the design is limited to three turbine stages. SOURCE: El-Masri (1985).

incidents of pollution shared through a common watershed or air basin now involve many nations, witness the concern for acid deposition in Europe and North America. What were once straightforward questions of conservation versus development now reflect complex linkages, witness the global feedbacks among energy and crop production, deforestation, and climate change that are evident in studies of the greenhouse effects. How these issues will effect the energy future is hard to predict.

One school of thought suggests that social structures and the ecological system possess a tremendous capacity to adapt to long-term changes. And because the uncertainty in scientific information is so great, the need or feasibility to take actions now is questioned. Another school

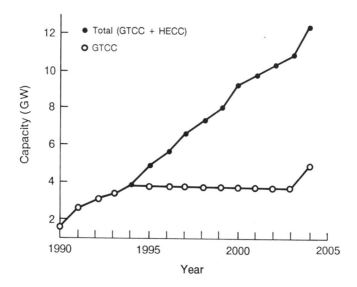

FIGURE 12  Projected capacity of gas turbine and high-efficiency combined-cycle (HECC) systems in the Northeast. SOURCE: Lee (1988, p. 143).

argues that most studies make convenient but unrealistic "surprise-free" assumptions regarding future developments in institutions, technology, and knowledge. Advocates of this perspective question whether the rate of climate change, under the assumption of continued, increasing emissions of infrared-trapping gases, would likely be too rapid to allow reasonable adaptive measures to be effective.

While the debate is going on, it is extremely interesting to note that over the past 100 years, the global primary energy system has moved progressively toward hydrogen-rich quality fuels, as shown by Figure 14. The hydrogen-to-carbon (H/C) ratio for fuelwood is roughly 0.1; for coal, 1.0; for oil, 2.0; and for natural gas, 4.0. The implications of this trend are far reaching, especially in light of recent discussions on the "airborne fraction" of emitted carbon dioxide (the portion of $CO_2$ emitted that remains in the air). The fact that the deep oceans are huge sinks for $CO_2$ instills hope that increases in the atmospheric concentration of $CO_2$ may be brought to zero without cutting the emission from fossil fuel combustion to zero. The question is how fast the upper mixing layer of the oceans can absorb $CO_2$ from the atmosphere. Firor (1988) recently suggested that "it is possible that society could come close to stabilizing the atmospheric burden of $CO_2$ with a 50 percent reduction in fossil fuel use." Although the quantitative conclusion will be a subject for debate for some time, there is general agreement that reduction in emissions from the supply side and

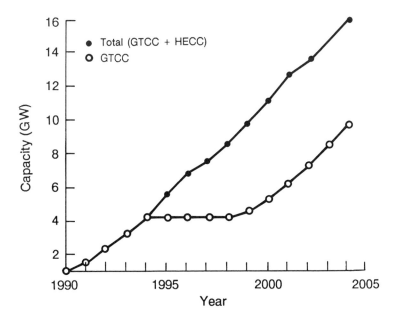

FIGURE 13  Projected capacity of high-efficiency combined-cycle (HECC) systems in the Southeast. SOURCE: Lee (1988, p. 143).

conservation on the demand side (improvement in efficiencies) are the right things to do.

More penetration by natural gas into the primary energy market is a step in that direction. It is gratifying that the historical trends are pointing that way. At the same time it is important to be mindful of the fact that substitution by natural gas may not be fast enough, but it will slow the buildup of $CO_2$. How effective that can be, we do not know.

We must also be aware that $CO_2$ is only one of the "greenhouse gases." Another important gas in that family is methane itself. Recently, rice fields have become known as an important source of methane. It is almost unimaginable that people will cut down rice consumption. Yet, non-methane-emitting rice production may be a suitable and very difficult challenge for technologists (biotechnologists) to work on: How can the quality of rice be maintained by a different process?

## BEYOND THE FOSSIL AGE

If natural gas is the bridge, what is on the other side of the river? There have been so many forecasts already that it is best not to add another. Forecasting in the energy field has proved a most hazardous profession.

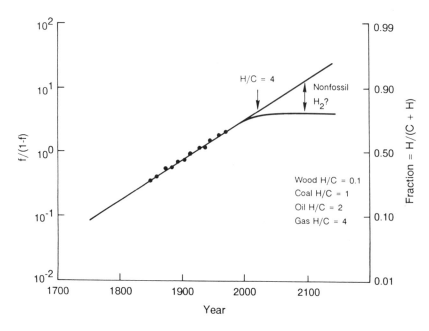

FIGURE 14 Evolution of the hydrogen-to-carbon ratio in primary energy sources, 1850–2100. SOURCE: Marchetti (1985).

Most forecasts have been wrong, and it is aptly suggested that the only way to forecast is to do it frequently! It is not productive to engage in debates on which forecast is right; rather, we should ask ourselves: If some of the forecast turns out to be right, what does it mean to us? Should we protect ourselves against particular events? These are only two of many questions planners should ask before they formulate a set of criteria for planning and design of future energy systems. For this purpose, a review of a few additional lessons from the past will help.

Perhaps the most important lesson is that uncertainty is a fact of life. Past belief in the single-trend forecast for oil prices has cost the United States billions of dollars in synfuel projects and in international bank loans. One criterion for future energy systems planning should be robustness against uncertainties.

The next important lesson is that the future will most likely not be simply a smooth extrapolation of the past but will be marked by fluctuations and new factors in competition. In business, unexpected events are referred to as contingencies. It is the responsibility of planners to imagine the surprises as best they can and then formulate a plan to deal with them, including triggering criteria and timing. Traditionally, studies of energy and ecological systems have been based on surprise-free models (Brooks, 1986). The gradual, incremental unfolding of the world system in such models with parameters derived from a combination of time series and cross-sectional analysis of the existing system is precisely why most forecasts are wrong. Thus, robustness must include contingency planning.

Another lesson is that defending the status quo may be a poor strategy. Societies are continually developing and seeking to meet new demands, be they in areas of safety or environmental quality. The responsibility of the technical community is to anticipate societal challenges and be ready with technological solutions.

Finally, over the long run, market economics is still the controlling factor. Neither experts nor the public should be misled by the power of noncommercial technical success and overestimate the power of government intervention. The United States believed that if it could send a man to the moon, it should be able to solve the energy crisis in the same way, by massive government financial support. However, the economics of the space race differs from that of the energy business. The United States believed it could change in a lasting way the economic attractiveness of technologies by building demonstration plants with massive government financial support. Looking back, we find what makes economic sense: in many areas the private sectors went ahead, without help from the government, to increase the efficiency of heat recovery in combustors, for example, and to add insulation and temperature controls in commercial or residential structures. For products that did not make sense, the "wise" organizations took government money for R&D, and the not so wise lost their own money in addition to that of the government. In the end, those options that did not make economic sense at the outset were never developed commercially.

These lessons suggest that future energy systems should at least meet the following criteria. They should be economic, efficient, safe, and of high quality. They should also be clean in relationship to the environment and robust with respect to uncertainties.

One point needs to be made with regard to the robustness requirement: that is, the system concept must be adaptive to a range of technological advances. One should neither count on revolutionary advances to the same degree as was widespread after 1973 nor seek only revolutionary advances. Future systems must build on the current technological menu and be ready to accept new items when they become available.

Energy engineers have been searching for a concept that is broadly applicable to energy systems planning and design, without being heavily constrained by issues such as indigenous supply; technological readiness; and local social, political, or economic conditions. From the technological

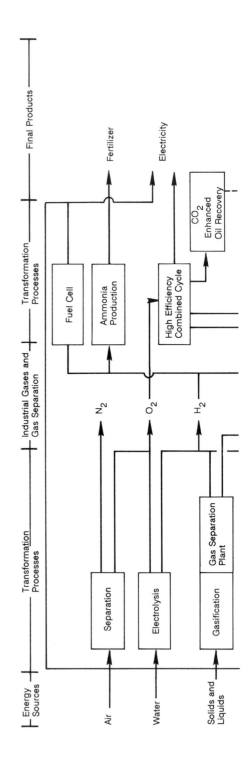

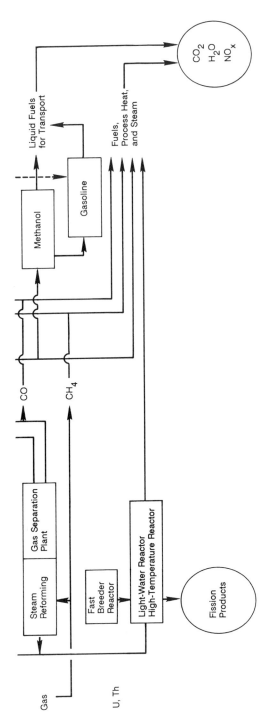

FIGURE 15  Functional diagram of integrated energy systems.

viewpoint, the concept should be evolutionary but adaptive to revolutionary technological changes. The integrated energy system is such a concept.

The concept of integrated energy systems is not new. The oil refinery/petrochemical complex and the steel mill are two good examples, even though they are never referred to as such. In an oil refinery and petrochemical complex, there is no clear distinction between product streams and energy streams. Crude oil, liquefied petroleum gases, natural gas, and other industrial gases are the primary materials used by the complex, but each is used for many purposes. For example, natural gas is used as fuel in heaters, as a feedstock, or as fuel for the unit making hydrogen. Industrial gases are exploited for their maximum benefit. The entire steam cycle is integrated: high-quality steam for turbines, medium-quality steam for process boilers, and low-quality steam for preheaters. Even more important, the steam system is integrated with the electricity system. The result is a robust, flexible system, highly efficient with respect to both energy and capital, and therefore economically sound. This is an integrated energy system, whether we call it that or not; neither would anyone design a modern refinery in any other way.

Another example is the conventional steel mill, for which the primary raw materials are coal and iron ore. Although the mill has a huge need for energy, it does not burn coal but instead uses coal as a chemical raw material. The coal is coked, a process that uses gases released when the coal is heated in the absence of air to remove the chemicals it contains. Coke is used in blast furnaces to reduce the iron oxide in the ore to metallic iron. Although one of the primary products of these processes is heat, the coal is not burned.

These are shining examples of integrated energy systems in which loss of heat or useful components is minimized, thereby enhancing economic efficiency. Operational and capital costs are also minimized. These systems are possible for several reasons: One is that the enterprise is big enough to rise above issues of investment and disciplinary (or professional) barriers. Another is that there are no significant regulations that stand in the way of their design, construction, and operation as integrated systems. These conditions do not hold true for all energy-related enterprises. Electric utility companies would have a variety of difficulties if they decided to enter the industrial gas business.

First, let us look at integrated energy systems (IES) from a conceptual point of view (Figure 15). On initial inspection, the IES appears to be a complex and unmanageable system. Although Figure 15 shows a number of boxes representing technological steps, the boxes represent options (for clarity, not all options are shown), not required components. The IES diagram offers alternatives in each stage of the system. There are five aspects to the system: (1) energy sources, including air and water;

(2) transformation processes (incoming fuels are transformed to industrial gases); (3) industrial gases and gas separation; (4) transformation processes (industrial gases are transformed to more usable energy forms, electricity or chemicals); and (5) product to final consumption.

The simplest integrated energy system, the cogeneration system using natural gas, for example, can be traced from gas to combined cycle to electricity and process heat and steam. Other systems can be constructed by selecting the appropriate options, as shown in Figure 16. The purpose here is not to promote any one alternative but to show the flexibility of the concept from three viewpoints: robustness with respect to uncertainty, ability to adapt to technological advances, and environmental protection.

The search for robustness must be an important part of strategic planning in energy. For a number of years, IIASA has conducted, in cooperation with Alan Manne of Stanford University, an annual survey of a number of forecasts of oil prices (Manne et al., 1985). Figure 17, containing recent survey results, shows that the range of all the forecasts is very wide and that the projected price of oil depends on the price at the time the forecasts were made. For planners, these survey results indicate that our knowledge of the future will always be uncertain. The transformation process between the primary sources and the set of intermediate industrial gases in Figure 15 provides robustness against uncertainty in supply. There are, for example, three potential sources of hydrogen: solid fuel such as coal, liquid fuel, or natural gas. There are two sources of carbon monoxide and oxygen. The energy required to produce these gases can come from hydrocarbons or from nuclear power. Of course, to make use of the robustness, the system must be designed to have the required flexibility.

The IES concept can also adapt to new technologies. Take fuel cells as an example. Today, the question of economic feasibility of fuel cells with phosphoric acid technology remains unanswered. With molten carbonate technology, the uncertainty lies mostly in the technical area. If the technical obstacles are overcome and the economics of either of the two technologies becomes attractive, fuel cells could be incorporated in a straightforward manner into an integrated energy system, as shown in Figure 15. The same can be said for high-temperature electrolysis, methanol production, new gas separation systems, renewable sources such as photovoltaics, and new nuclear reactor technologies, be they high-temperature gas reactors or fusion. The question is where, not whether, they belong in the system. Thus, the concept of integrated energy systems is friendly to both evolutionary and revolutionary technologies.

The ultimate dream for energy systems—zero emissions—can be accomplished only with a hydrogen economy. Integrated energy systems offer a technological road map toward this environmental goal. At present, all

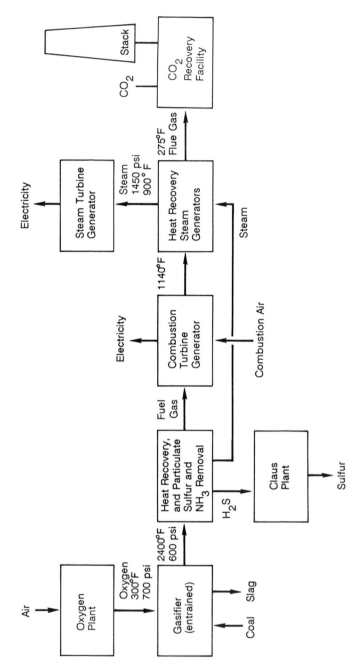

FIGURE 16 Integrated gasifier combined cycle with $CO_2$ recovery.

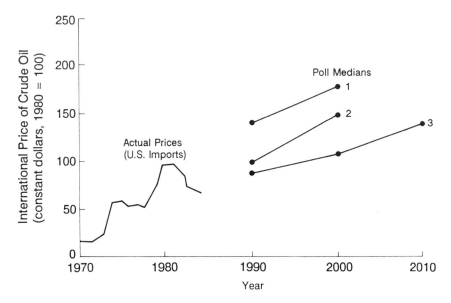

FIGURE 17 Changing outlook for oil prices: actual prices from 1970 to 1985 and median projections as published by the IIASA International Energy Workshop, 1981–1985. Polls on expected price of crude oil were taken at roughly 18-month intervals: poll 1 in December 1981, poll 2 in July 1983, and poll 3 in January 1985. SOURCE: Manne et al. (1985).

parties in the energy production chain share responsibility for environmental protection. In an integrated energy system, the responsibility is focused on a well-identified set of processes (the left-hand portion of Figure 15). This should make the job of environmental protection easier. Residential, commercial, and industrial consumers would welcome energy forms with greatly reduced environmental concerns.

Implementation of integrated energy systems requires major changes in industrial infrastructure. It requires integration of chemical, petrochemical, and electric power industries, along with regulatory and economic adjustments. How fast trends may advance is unknown, but some indications are encouraging. Carbon dioxide from power plants is used for enhanced recovery of oil. A new energy project outside of Stockholm is based on the integrated concept. The role of engineers is to have the technologies ready to meet social demand. In the energy area, it appears that technologists are ready with timely and environmentally attractive proposals to move to advanced fossil fuel systems and beyond.

# REFERENCES

Adelman, M. A. 1987. Are We Heading Towards Another Energy Crisis? Paper presented to Oil Policy Seminar of the Petroleum Industry Research Foundation, Washington, D.C., September 29, 1987.

Brooks, H. 1986. The typology of surprises in technology, institutions, and development. Pp. 325–350 in Sustainable Development of the Biosphere. W. C. Clark and R. E. Mann, eds. New York: Cambridge University Press.

El-Masri, M. A. 1985. On thermodynamics of gas turbine cycles. ASME Transactions 107:880–889.

Firor, J. 1988. Pp. 103–105 in Climatic Change, Vol. 12. Boston: Kluwer Academic Publishers.

Grübler, A., and N. Nakicenovic. 1988. The dynamic evolution of methane technologies. Pp. 13–14 in The Methane Age, T. H. Lee, H. R. Linden, D. A. Dreyfus, and T. Vasko, eds. Boston: Kluwer Academic Publishers.

Lee, T. H. 1988. Combined cycle systems: Technology and implications. Pp. 131–145 in The Methane Age, T. H. Lee, H. R. Linden, D. A. Dreyfus, and T. Vasko, eds. Boston: Kluwer Academic Publishers.

Lee, T. H., and N. Nakicenovic. 1988. Technology life-cycles and business decisions. International Journal of Technology Management 3(4):411–426.

Lee, T. H., H. R. Linden, D. A. Dreyfus, and T. Vasko, eds. 1988. The Methane Age. Boston: Kluwer Academic Publishers.

Manne, A. S., L. Schrattenholzer, A. N. Svoronos, and J. L. Rowley. 1985. International Energy Workshop 1985. Part I: Summary of Poll Responses. Laxenburg, Austria: International Institute for Applied Systems Analysis.

Marchetti, C. 1985. When will hydrogen come? International Journal of Hydrogen Energy 10:215.

Marchetti, C., and N. Nakicenovic. 1979. The dynamics of energy systems and the logistic substitution model. International Institute for Applied Systems Analysis. Report RR-79-13. Laxenburg, Austria.

Rogner, H.-H. 1988. Natural gas and technical change: Results of current gas studies. Pp. 61–84 in The Methane Age, T. H. Lee, H. R. Linden, D. A. Dreyfus, and T. Vasko, eds. Boston: Kluwer Academic Publishers.

Tabors, R. D., and D. P. Flagg. 1986. Natural gas fired combined cycle generators: Dominant solutions in capacity planning. IEEE Transactions on Power Systems PWRS-1(2):122–127.

Tilton, J. E. 1984. Material substitution: Lessons from the tin-using Industry. International Institute for Applied Systems Analysis. Report RR-84-009. Laxenburg, Austria. Reprinted from Material Substitution: Lessons from the Tin-Using Industry, J. Tilton, ed. Washington, D.C.: Resources for the Future, 1983.

Tilton, J. E., and H. H. Landsberg. 1984. Nonfuel minerals: The fear of shortages and the search for policies. International Institute for Applied Systems Analysis. Report RR-84-008. Laxenburg, Austria. Reprinted from U.S. Interests and Global Natural Resources: Energy, Minerals, Food, E. N. Castle and K. A. Price, eds. Washington, D.C.: Resources for the Future, 1983.

# Protecting the Ozone Layer: A Perspective from Industry

JOSEPH P. GLAS

Protection of the ozone layer is a model of the way science, technology, and public policy can work together to achieve global agreement and action. The progress to date is a result of three basic factors: a shared goal of protecting the environment, fundamental agreement on the science, and advances in technology to meet societal needs.

## ORIGINS OF CONCERN

In the more than 15 years since chlorofluorocarbons (CFCs) were first implicated in possible ozone depletion, those industries producing and using CFCs have asserted that policy should be based on the best available scientific information.[1] As a company, Du Pont, the world's largest producer, has sought to support and pursue development of the science, to base its position on the best available science, and once established, to act aggressively on its position.

Clearly, the attention paid to this issue over the past decade and a half is a product of science. Lovelock's invention in 1970 of the electron capture detector for gas chromatography first provided the capability of measuring CFCs in the atmosphere in parts per trillion. By revealing that

---

This chapter is based on a talk given at the National Academy of Engineering Annual Meeting, September 29, 1988. It includes supporting information provided with the assistance of the National Academy of Engineering Program Office.

CFCs were accumulating in the atmosphere, Lovelock's measurements in the early 1970s indirectly provided the first evidence for possible concern about these compounds (Lovelock, 1971).

Du Pont's reaction to the information was to arrange a seminar on "The Ecology of Fluorocarbons" for the world's CFC producers. The year was 1972. The invitation from Raymond Mc Carthy, then research director of Freon Products, previewed future industry responses:

> Fluorocarbons are intentionally or accidentally vented to the atmosphere worldwide at a rate approaching one billion pounds per year. These compounds may be either accumulating in the atmosphere or returning to the surface, land or sea, in the pure form or as decomposition products. Under any of these alternatives, it is prudent that we investigate any effects which the compounds may produce on plants or animals now or in the future.

As a result of that industry symposium, a research program was established to investigate the ultimate fate and impact of CFCs in the atmosphere. Nineteen companies formed the Chemical Manufacturers Association's (CMA) Fluorocarbon Program Panel, a group that has funded well over $20 million in research to date at academic and government facilities worldwide, including support of recent Antarctic expeditions.

In 1974, about two years after the industry symposium and initiation of the enhanced research program, Molina and Rowland (1974) published an article proposing that the ultimate fate of CFCs was ultraviolet photodecomposition in the stratosphere with the release of chlorine atoms. Through a series of rapid chemical reactions, these chlorine atoms might cause a reduction in the total amount of stratospheric ozone (Figure 1). The concerns of these and other scientists led the industry group to redirect its research activities toward confirming or refuting the initial conclusion regarding stratospheric photolysis of CFCs and the possible impacts of that decomposition, including potential ozone depletion.[2]

Stratospheric science was in its infancy at the time. There was no reliable means of checking the validity of the ozone depletion theory. Led by government funding agencies, but with significant input from industry, scientists from government, academia, and industry undertook the enormous task of developing the scientific base, including a greatly expanded worldwide set of measurements, with the goal of predicting future ozone amounts. One of the results was the development of more realistic and comprehensive models which, by the early 1980s, were used by policymakers to study potential regulatory scenarios.

Despite shortcomings in the amount and quality of data, the science of the late 1970s told us three things. First, the time scales involved are long for both the onset and the decay of any effects from CFCs (Figure 2). Although available evidence indicated that there appeared to be sufficient time to perform research to reduce uncertainties, control measures would

**Production**

$$O_2 + \text{Solar UV } (\lambda < 220 \text{ nm}) \longrightarrow 2O$$
$$2(O + O_2 + M) \longrightarrow O_3 + M)$$

Net: $3O_2 \longrightarrow 2O_3$

**Destruction**

$$X + O_3 \longrightarrow XO + O_2$$
$$O_3 + \text{Solar UV } (\lambda < 310 \text{ nm}) \longrightarrow O_2 + O$$
$$XO + O \longrightarrow X + O_2$$

Net: $2O_3 \longrightarrow 3O_2$

$X = NO, OH, Cl, O_2$

FIGURE 1 Production and destruction of ozone. Ozone is produced and destroyed naturally at the rate of about 300 million tons per day. Production occurs primarily as the result of dissociation of molecular oxygen by absorption of solar ultraviolet radiation. Oxygen molecules can also combine with oxygen atoms to form ozone, if a suitable liquid or solid surface (M) is present. Ozone is destroyed by several natural catalytic cycles. About 70 percent of the natural destruction is due to the nitrogen cycle. Chlorine is believed to be the principal agent upsetting the natural balance of ozone production and destruction. There is concern that increasing concentrations of CFCs could add enough chlorine to the atmosphere to increase the net destruction rate and decrease the net amount of ozone. SOURCE: Du Pont Company.

probably be required well in advance of any identifiable damage to the biosphere.

Second, the science involved is incredibly complex, with relevant new chemical reactions being discovered regularly, and in key respects is unproven. Scientists, government, and industry were all mindful of attempts to predict stratospheric ozone destruction by nitrogen oxide emissions from supersonic transport planes (SSTs) in the early 1970s, and how results had shifted dramatically (Figure 3).

Third, the processes and effects are clearly global. Because any CFCs entering the atmosphere would be mixed throughout the atmosphere relatively quickly, no individual geographic region had exclusive control over its own ozone layer. Moreover, CFCs were consumed in significant amounts in many nations and regions (Figure 4). The quick conclusion was that if there were a problem, the entire world would have to act in near unison.

Du Pont's corporate environmental policy, formulated in the late 1930s,

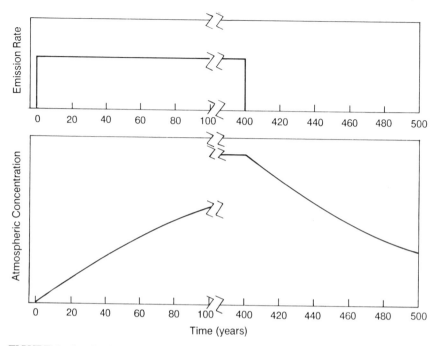

FIGURE 2 Implications of long atmospheric lifetime of CFCs. Both the emission rate and the concentration are plotted as arbitrarily chosen, linear scales. The concentration responds slowly to change in the emission rate. SOURCE: Du Pont Company.

commits Du Pont to "determine that each product can be made, used, handled and disposed of safely and consistent with appropriate safety, health and environmental quality criteria." In fact, in 1975, Chairman of the Board Irving Shapiro stated publicly that if there were credible scientific evidence of harm to human health or the environment, Du Pont would cease manufacture of fully halogenated CFCs. About once a year, dating back to the mid-1970s, Du Pont formally reviewed its position. The question was always the same: On the basis of what we know from the science, what—if any—controls are appropriate? Once the company's position on controls had been determined, consideration was given to implications for business strategies.

## ROLE OF TECHNOLOGY

Inevitably, technology became a key aspect of the ozone issue. CFCs had been invented around 1930 as a safe alternative to ammonia and sulfur dioxide for use in home refrigerators (see Friedlander, this volume). The intent was to eliminate the toxicity, flammability, and corrosion concerns of other chemicals by developing a stable chemical with the right thermodynamic properties. That effort was so successful that the new compounds were also quite easy to make and rather inexpensive. New applications for a safe class of chemicals with the properties of CFCs were plentiful and the market blossomed. Currently, virtually all refrigeration, commercial air conditioning, defense and communications electronics, medical devices, and high-efficiency insulation use CFCs in some way.

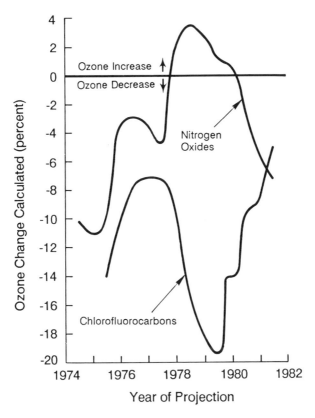

FIGURE 3 Long-term stratospheric ozone change projections from constant emission rates. Long-term projections of stratospheric ozone change, based on constant emission rates, provide an example of how complex, poorly understood processes can significantly affect the predictions of a mathematical model of man-made environmental changes. This graph shows stratospheric ozone change estimates from a series of models developed to predict the effects in the next century of steady-state emissions of both CFCs and nitrogen oxides from a hypothetical fleet of SSTs. The calculations were made over a number of years at Lawrence Livermore National Laboratory. SOURCE: Schneider and Thompson (1985).

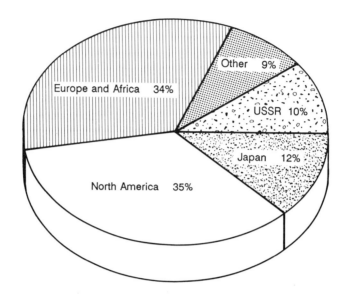

FIGURE 4 Approximate consumption of CFCs by country or region, 1988. Total world consumption was 2,510 million pounds. SOURCE: Du Pont Company estimates.

Today, some 50 years after the development of CFCs, we have redefined "safe" to mean something not quite so stable—that is, not as stable in the atmosphere—which still retains the desirable properties afforded by stability during use.

## PRECAUTIONARY ACTIONS

In the mid-1970s, despite limited scientific understanding and evidence, several environmental groups insisted that precautionary action be taken to control CFCs. They focused attention on the so-called nonessential uses of CFCs, primarily aerosol propellants. Despite strong protests from industry, a few countries, led by the United States, banned those uses in 1978. In my view, because this unilateral action was not based on unequivocal scientific guidance, the ultimate result was broader global inaction for almost 10 more years.

Anticipating a potential need for substitutes if regulations were promulgated, Du Pont initiated a large research effort in the mid-1970s to identify and, if possible, develop alternative chemicals to replace the fully halogenated CFCs. In 1980, after numerous candidates had been rejected as too toxic, significantly more costly to manufacture, or not usable in their intended applications, Du Pont published its conclusions about the most promising candidates (Du Pont Company, 1980).

As long as the existing products were freely available, the new candidates, being less cost-effective, could never hope to compete unless some external factor drove market demand. Regulations that could create demand for alternatives were not forthcoming. Additionally, the U.S. ban on aerosols—a segment that accounted for about 50 percent of the U.S. CFC market—had forestalled CFC growth sufficiently so that additional controls seemed unwarranted and unlikely (Figure 5).

Although advances in science had led to numerous refinements in model projections of future ozone levels, significant uncertainties remained in the early 1980s. At the same time, published analyses of atmospheric measurements indicated no persistent trend in total column ozone (Table 1). This supported the belief that there would not be significant changes in ozone in the near term.

However, continuing uncertainties led to renewed interest in regulation of CFCs. Anticipating such action, in the mid-1980s CFC producers and users formed the Alliance for Responsible CFC Policy. The expressed purpose of the Alliance is to advocate that policies be based on the best science and that only a global approach to controls would be effective in protecting the ozone layer.

In October 1980, reacting to model calculations that ozone depletion might reach 15–20 percent at the end of the next century, the Environmental Protection Agency (EPA) published an Advance Notice of Proposed Rulemaking (ANPR). The ANPR suggested the need for additional controls and an eventual phaseout of CFC production and use. Subsequent model results, combined with recognition of trends in atmospheric concentrations of other trace gases, indicated that net changes in ozone, if any, were likely to be insignificant, provided there was no large growth in CFC production (National Research Council, 1982). This again removed support for regulation. The ANPR was left open, with no decision by the EPA to either pursue or abandon it. After it became apparent that there would be no controls to drive demand for substitute products, Du Pont curtailed its R&D efforts on alternatives.

## INTERNATIONAL EFFORTS

International attention had remained focused on the ozone issue through the United Nations Environment Program (UNEP) which, in 1977, organized the Coordinating Committee on the Ozone Layer, that met at least biennially and published a series of scientific assessments. Responding to the concerns expressed in those reports, in 1981 UNEP formed an ad hoc group to consider development of a global convention for protection of the ozone layer. After unsuccessful attempts to negotiate a convention that would include provisions aimed at control of CFCs, the group abandoned

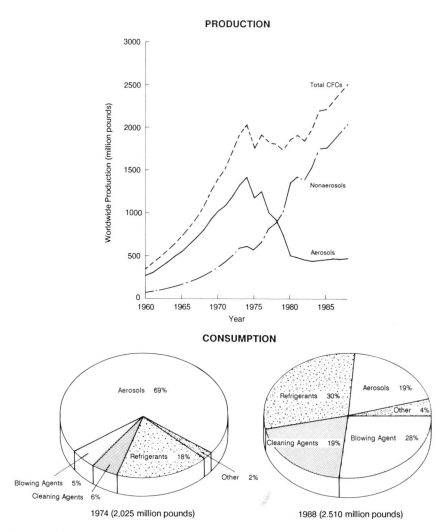

FIGURE 5 Worldwide production and consumption of CFCs. *Above*, estimated worldwide total production of CFCs for both aerosol and nonaerosol use from 1960 to 1988, *below*, differences in consumption by application in 1974 and 1988. Although the United States banned the use of CFCs as aerosol propellants for most applications in 1978, many countries did not. SOURCE: Du Pont Company.

that effort and proceeded with a framework convention calling for global cooperation on research, data collection, and technology exchange.

The UNEP Vienna Convention for the Protection of the Ozone Layer was adopted in March 1985. The convention was designed so that protocols could be added requiring specific control measures. The group also outlined

TABLE 1 Trends in Total Ozone Change, as Reported in the Early 1980s

| Period | Change (percent) | | Reference |
|---|---|---|---|
| 1970-1978 | +0.28 | ± 0.67 | Reinsel et al. (1981) |
| 1970-1979 | +1.5 | ± 0.5 | St. John et al. (1982) |
| 1970-1979 | +0.1 | ± 0.55 | Bloomfield et al. (1983) |
| 1979-1983 | -0.003 | ± 1.12 per decade | Reinsel et al. (1984) |
| | (-0.14 | ± 1.08) per decade with sunspot series in model | |

SOURCE: World Meteorological Organization-National Aeronautics and Space Administration (1986).

plans for a series of workshops to evaluate further the need for such controls and explore possible means of control that could find worldwide acceptance.

Concurrent with these regulatory discussions, a worldwide group of experts was engaged in a comprehensive review of the science. Completed in late 1985, the study concluded that there was no evidence of global ozone depletion and forecast no depletion based on limited growth in CFC usage (WMO-NASA, 1986). However, model calculations that assumed sustained growth in CFC emissions did predict depletion in ozone (see Figure 6).

Just as the study was being completed, British scientists uncovered the first evidence of significant but temporary changes in ozone over Antarctica (Farman et al., 1985). Despite the lack of consensus about causes of the so-called Antarctic hole, the observation of real change again focused world attention on the issue of CFCs and their effects on stratospheric ozone.

## RENEWED CONTROL EFFORTS AND INDUSTRY LEADERSHIP

While progress was being achieved at the international level, in the United States the Natural Resources Defense Council (NRDC) filed suit against the EPA. The NRDC claimed that by not following up the 1980 ANPR with a decision regarding future regulations, the EPA had failed to meet its obligations under the Clean Air Act. The suit was settled late in 1985 with the publication of EPA's Stratospheric Ozone Protection Plan, which called for a series of U.S. workshops to be held in conjunction with those planned by UNEP. They were to be followed by an EPA decision by May 1, 1987, and publication of a final rule, if needed, by November 1, 1987.

Through this period, the pattern of CFC use by industry had begun to change. By the mid-1980s, the growth of refrigeration, cleaning agents, and foam insulation markets more than offset the decline of CFCs in aerosol

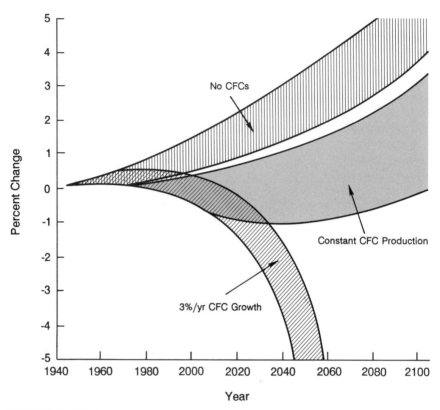

FIGURE 6 Calculated ozone change over time. The range of changes in total ozone calculated by the various modeling groups from the United States and Europe is shown for three assumptions for past and future consumption of CFCs. The top range shows calculated changes if CFCs were never emitted to the atmosphere. The middle shows calculated changes if historical CFC consumption rates are assumed through 1985 and constant consumption at the 1985 rate thereafter. The bottom range shows calculated changes if historical CFC consumption rates through 1985, with compounded growth of the consumption rate at 3 percent per year thereafter. Ozone amounts are calculated to increase in the top and middle ranges because of the effects of increasing amounts of carbon dioxide and methane in the atmosphere. SOURCE: Data were assembled from a variety of sources including WMO-NASA (1986).

markets in the United States and Canada (see Figure 5). Furthermore, forecasts projected continued growth in demand, due in large part to the expectation that developing countries would want the services provided by CFCs.

These growth forecasts, coupled with computer model predictions of ozone depletion if there were sustained growth in CFC emissions, once again increased concerns (see Figure 6). With the body of information acquired over the previous decade, it became clear that, regardless of quantitative results, significantly increased emissions were likely to result in decreases in ozone. Based on this information, the worldwide CFC industry, led in September 1986 by Du Pont and the Alliance for Responsible CFC Policy, first advocated international efforts to limit long-term growth of CFC emissions.

The new policy argued that controls should be global and should focus on net worldwide emissions to the stratosphere rather than on individual uses or countries. The failure of the 1978 U.S. aerosol ban to halt worldwide growth was cited as an example of the inability of such isolated actions to have lasting effects.

It is difficult to say whether any specific factor led to Du Pont's 1986 policy change. Probably most influential was growing confidence in the models' ability to predict ozone depletion for growth scenarios, coupled with recognition that demand for CFCs was growing at a significant rate and would likely continue to grow if left alone. In 1986 Du Pont also reactivated research on chemical substitutes; the reasoning was that alternatives would eventually be needed, regardless of cost.

## A HISTORIC INTERNATIONAL AGREEMENT

The announcements by U.S. industry in 1986 contributed significantly to productive international negotiations that began in December of that year. Du Pont was an active participant throughout, as was the Alliance for Responsible CFC Policy. With some initial reluctance, other leading CFC producers also offered their support for an international agreement. The basis for consensus was a shared goal of protecting the environment, commitment to active participation in efforts to advance scientific understanding, and agreement that any regulations should be based on sound information. The growing industry support led negotiators to a productive discussion of the implications of different regulatory proposals.

Although industry participated in the discussion of various control strategies, it pointed out that technical analyses had demonstrated only the need for limitations to *growth* in CFC emissions. Some environmental groups, on the other hand, insisted that if there were indeed any level of emissions that was unsafe, and that level could not be determined accurately, then the only appropriate action was elimination of all CFC emissions.

The results of these developments were twofold. First, the search for a structure for the proposed regulations became a complex interplay of national economic interests seeking a straightforward yet equitable solution. Second, the stringency and timing of the regulations became a political struggle between supporters of aggressive controls, on one side, and those who sought a more cautious approach, on the other. A sound scientific base indicating the need for some level of controls maintained the discussions.

From the standpoint of industry, if the negotiators could develop regulations that CFC producers and users worldwide could meet without severe economic costs and safety risks, then the process would clearly advance. Much of industry had already accepted that there should be some kind of limit. This acceptance contributed to the development of the international process and helped government negotiators to focus on the issues necessary to gain a consensus.

Ensuing negotiations in the late spring and early summer of 1987 led to signing of the Montreal Protocol in mid-September (UNEP, 1987). It dealt with a broad range of considerations. This protocol had to determine a "safe" level of emissions. It had to be acceptable to developing countries, who were seeking the economic and societal benefits that CFCs had made possible for developed countries. Another important consideration was to maintain free-flowing international trade in what had become a truly global market. Most important, the protocol had to be a living document. There was a need for sufficient flexibility to adjust the terms of the protocol in response to scientific, technological, and socioeconomic developments. (The box on page 149 summarizes the provisions of the Montreal Protocol.)

As the negotiations were nearing completion, it became apparent to Du Pont and others that the need for alternative compounds would likely arise sooner than expected. The search began anew for ways to reduce the time needed for development.

One clear need was a way to speed up initiation of the six to seven years of toxicity testing normally required for such high-volume chemicals. Du Pont contacted other producers who had publicly expressed interest in developing alternatives. A core group then identified the most promising products and concluded that a cooperative effort would generate the needed toxicity information most efficiently. An invitation was then extended to all other CFC-producing companies. By January 1988, the 14-member Panel for Alternative Fluorocarbon Toxicity Testing was formed and an aggressive five-year program was under way.

## CREDIBLE SCIENTIFIC EVIDENCE

The ink on the Montreal agreement (UNEP, 1987) was barely dry and the ratification process had just begun when, on March 15, 1988, NASA's Ozone Trends Panel (Watson et al., 1988) announced new findings that raised serious questions about whether the restrictions on CFC production and use contained in the protocol were adequate to protect stratospheric ozone. Figure 7 shows the 1987 Antarctic ozone "hole" that was the central motivating finding in the new assessment.

The Montreal Protocol is designed to help reach international agreement on control of the production and consumption of certain chlorofluorocarbon and halon compounds. For developed countries, it calls for a freeze in CFC-11, 12, 113, 114, and 115 at 1986 consumption levels in mid-1989, with a 20 percent reduction from 1986 levels in mid-1993, and a 50 percent reduction by July 1, 1998. Halon-1211, 1301, and 2402 would be frozen at 1986 consumption levels in 1992, or three years after the protocol became effective.

The Montreal Protocol required ratification by nations representing at least two-thirds of total world consumption of CFCs and halons. The protocol entered into force on January 1, 1989.

| Montreal Protocol Participants | |
|---|---|
| Argentina | Maldives |
| Australia | *Malta |
| Austria | *Mexico |
| *Belgium | Morocco |
| Burkina Faso | *Netherlands |
| *Byelorussian SSR | *New Zealand |
| *Canada | *Nigeria |
| Chile | *Norway |
| Congo | Panama |
| *Denmark | Philippines |
| *EEC | *Portugal |
| *Egypt | Senegal |
| *Federal Republic of Germany | *Singapore |
| *Finland | *Spain |
| *France | *Sweden |
| Ghana | *Switzerland |
| *Greece | Thailand |
| Indonesia | Togo |
| *Ireland | *Uganda |
| Israel | *Ukrainian SSR |
| *Italy | *United Kingdom |
| *Japan | *United States |
| *Kenya | *USSR |
| *Luxembourg | Venezuela |

*Ratified: 46 signatories, 31 ratifiers, January 9, 1989.

Within three days of the Ozone Trends Panel report, internal discussions on the findings reached Du Pont's Executive Committee; after discussing the new findings with company scientists, the committee immediately decided to adopt a new position. Less than a week later, on March 24, Du Pont publicly announced its goal of an orderly transition to the phaseout of production of fully halogenated CFCs and the introduction of alternative chemicals and technologies as an essential part of the phaseout. The company also reiterated support for the Montreal agreement as the only effective means of addressing the issue on a global basis and called for a strengthening of the protocol to consider further global limitations on the emissions of CFCs.

Since the announcement, CFC producers such as Pennwalt Corporation, Allied-Signal, and Imperial Chemical Industries, as well as the Alliance for Responsible CFC Policy, the Food Service and Packaging Institute, the American Refrigeration Institute, and several CFC users have either taken steps to reduce the use of CFCs or urged more stringent controls through the international process.

Following the phaseout decision, Du Pont again reviewed the aggressiveness of its alternative R&D efforts to ensure that every possible measure was being taken to accelerate the program. Greater financial risks were to be taken, but safety considerations were not to be compromised. As a result of this review, numerous additional initiatives have been undertaken, especially in the area of applications development.

Du Pont's goal is to phase out its production of CFCs as soon as possible. The target is to complete the phaseout not later than the end of the century. Six operations are dedicated to developing alternatives, including four pilot plants, a small-lot production facility, and a commercial-scale plant.

In September 1988 Du Pont announced plans to invest more than $25 million in the world's first commercial-scale plant to produce HFC-134a, the leading candidate to replace CFC-12 in the largest U.S. market segment—refrigeration and air conditioning. This plant will be located in Corpus Christi, Texas, and will have the capability to expand to a much larger-scale facility in the future. In 1988 Du Pont spent more than $30 million for process development, market research, applications testing, and small-lot production of CFC alternatives; it expects to spend more than $45 million for R&D in 1989.

Our plan at Du Pont is to commercialize a series of new products during a three- to five-year period beginning in 1990. This schedule assumes favorable toxicology, process development and plant design, a favorable business climate, and reasonable financial risks. If problems arise in any aspect of the commercialization process, the schedule for new products will have to be reevaluated.

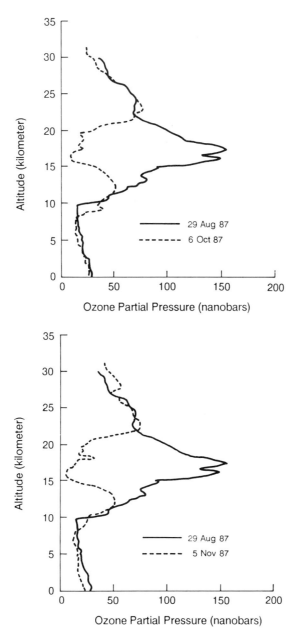

FIGURE 7 Vertical profiles of ozone using electrochemical ozonesondes from McMurdo Station in Antarctica, August–November 1987. The figures show the drop from Antarctic winter (August) to unusually low levels in Antarctic spring (October–November). By October the total ozone over Antarctica had been reduced by more than 50 percent of its 1979 value. Local depletion was as great as 95 percent at altitudes of 15–20 kilometers. SOURCE: Watson (1989, p. 19).

Du Pont's programs will be inadequate in the long term without global application and cooperation. Du Pont and all other firms must continue to believe in and support the international process established with the Montreal Protocol, hoping that all nations can, in fact, work together to stengthen the protocol to achieve a timely global phaseout. Figure 8 shows the implications for CFC concentrations for a range of emission scenarios.

In the United States alone, there is now more than $135 billion worth of installed equipment dependent on current CFC products. Virtually all of this equipment, some of it with a remaining useful lifetime of 20 to 40 years, could require replacement or modifications. For some industries, the impact of change will be even more dramatic. Entire industries could fold and, perhaps, be replaced by others.

Whatever action is taken, and whenever it occurs, technology will continue to play a critical role. The rate of technological progress and the degree of risk are inextricably related. In the extreme, a ban on CFCs before alternative chemicals or technologies can be put into place would mean lapses in the distribution of blood, other medical supplies, and up to 75 percent of the U.S. food supply. It could also force shutdowns of many modern office buildings that require air conditioning, as well as many U.S. manufacturing operations.

From a CFC standpoint, what action would appear to be most beneficial to the ozone layer? In the absence of scientific certainty, but based on the best available science, the prudent answer is a virtual phaseout of the suspect CFCs. Then the question is, What are the costs and risks associated with such a decision? If society is forced to choose a toxic or flammable, but legally allowed, chemical for refrigeration as the only alternative available to prevent critical shortages, it will be committed to a known risk in the home and workplace rather than a less certain global risk.

A final critical question deals with global concerns. What mechanism can be used to ensure that unified action is taken on a global scale? History has shown that less environmentally conscious governments are ready to let the United States take the more aggressive actions to enhance environmental protection. In today's world economy, competitive advantages are sought wherever they can be found. A simplistic policy approach based on the premise that "what is obvious to me must be best for everyone" is doomed to failure.

## CONCLUSION

A lot has been learned about the science of stratospheric ozone in the nearly 20 years since Lovelock's early work in his basement laboratory. More important, through efforts to address the ozone depletion issue, we appear finally to have found a way to behave as a global community and

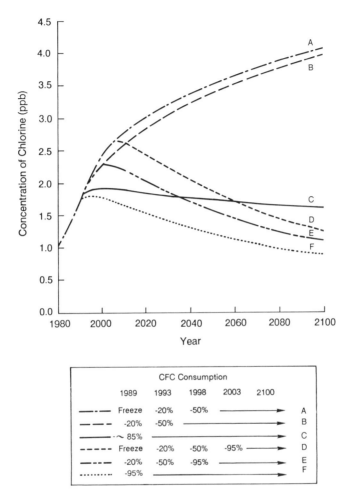

FIGURE 8 Effect of CFC reduction, showing total amount of calculated chlorine in the atmosphere from CFCs for several assumptions of future global use rates. There is very little difference between the two cases (A and B) in which CFC emissions are not decreased by more than the 50 percent reduction required by the Montreal Protocol. The effect of moving forward each reduction step by one control period is minimal (B). A reduction by 80–85 percent (C) maintains the atmospheric levels of chlorine from CFC emissions at an almost constant level. Adding a 95 percent use-reduction step (D) to the protocol results in reductions in the contribution of chlorine from CFCs. Over the next century, it would decrease by 75 percent the chlorine that would be added to the atmosphere if the protocol is not modified. Accelerating the reductions (E) has a relatively small effect, in part because other compounds contribute about 1.6 parts per billion (ppb) of chlorine to the atmosphere. A 95 percent reduction in 1989 (F) leads to chlorine decreases that begin almost immediately. However, such a reduction is not practical in view of the amount of CFCs required to meet basic societal needs, including refrigeration of food and medical supplies.

make a commitment to reduce the overall risks to society in the future. We have learned that it is possible to act quickly and forcefully by building on a common goal of protecting the environment and on fundamental agreement in science. The development of new technologies has provided what appear to be viable options for meeting society's needs.

Today's visible results are only the beginning of what will, I believe, become a major success story in environmental protection. Industrial firms will continue to take a strong leadership role in helping to bring about a global solution to this global environmental issue—an issue that should be a prototype for dealing with other global issues such as the greenhouse effect.

## NOTES

1. For reviews of scientific aspects of the ozone question, see Garfield (1988), National Research Council (1989), and Rowland (1989).
2. Concern for the protective ozone layer around the world stems from the fact that this layer, primarily 10–20 kilometers above the earth, screens out most of the biologically damaging ultraviolet radiation emitted by the sun (Maugh, 1980).

## REFERENCES

Bloomfield, P., G. Oehlert, M. L. Thompson, and S. Zeger. 1983. A frequency domain analysis of trends in Dobson total ozone records. Journal of Geophysical Research 88:8512–8522.

Du Pont Company. 1980. Fluorocarbon/Ozone Update. Wilmington, Del.: E. I. du Pont de Nemours and Company.

Farman, J. C., G. B. Gardiner, and J. D. Shanklin. 1985. Large losses of total ozone in Antarctica reveal seasonal $ClO_x/NO_x$ interaction. Nature 315:207–210.

Garfield, E. 1988. Ozone layer depletion: Its consequences, the causal debate, and international cooperation. Current Contents (6):3–13.

Lovelock, J. E. 1971. Atmospheric fluorine compounds as indicators of air movements. Nature 230(April 9):379.

Maugh, T. H. 1980. Ozone depletion would have dire effects. Science 207:394–395.

Molina, M., and F. S. Rowland. 1974. Stratospheric sink for chlorofluoromethanes: Chlorine atom catalyzed destruction of ozone. Nature 249:810–812.

National Research Council. 1982. Causes and Effects of Stratospheric Ozone Reduction: An Overview. Environmental Studies Board, Commission on Natural Resources. Washington, D.C.: National Academy Press.

National Research Council. 1989. Ozone Depletion, Greenhouse Gases, and Climate Change. Washington, D.C.: National Academy Press.

Reinsel, G., G. C. Tiao, M. N. Wang, R. Lewis, and D. Nychka. 1981. Statistical analysis of stratospheric ozone data for the detection of trend. Atmospheric Environment 15:1569–1577.

Reinsel, G., G. C. Tiao, J. L. DeLuisi, C. L. Mateer, A. J. Miller, and J. E. Frederick. 1984. Analysis of upper stratospheric Umkehr ozone profile data for trends and the effects of stratospheric aerosols. Journal of Geophysical Research 89:4833–4840.

Rowland, S. F. 1989. Chlorofluorocarbons and the depletion of stratospheric ozone. American Scientist 77(January–February):36–45.

Schneider, S. H., and S. L. Thompson. 1985. Future changes in the atmosphere. Pp. 397–430 in The Global Possible, R. Repetto, ed. New Haven, Conn.: Yale University Press.

St. John, D., W. H. Bailey, W. H. Fellner, J. M. Minor, and R. D. Sull. 1982. Time series analysis of stratospheric ozone. Commun. Stat., Part A 11:1293–1333.

United Nations Environment Program (UNEP). September 16, 1987. Montreal Protocol on Substances That Deplete the Ozone Layer. Montreal: UNEP.

Watson, R. T., M. J. Prather, and M. J. Kurylo. 1988. Present state of knowledge of the upper atmosphere 1988: An assessment report. NASA Reference Publication No. 1208. Washington, D.C.: National Aeronautics and Space Administration.

World Meterological Organization-National Aeronautics and Space Administration (WMO-NASA). 1986. Atmospheric ozone 1985: Assessment of our understanding of the processes controlling its present distribution and Change. Global Ozone Research and Monitoring Project, Report No. 16, 3 vols. Geneva: WMO.

# 3
# Social and Institutional Aspects

# The Rise and Fall of Environmental Expertise

VICTORIA J. TSCHINKEL

Various professions have thrust themselves forward with enthusiasm, pride, and touching self-confidence as the key to saving and managing our natural environment. Physicians, engineers, biologists, and lawyers have all contributed their talents and prejudices to the cause. Given their differences of view, it is probably not surprising that we are where we are today. Thirty years after the birth of the modern environmental movement, we are still questioning what the real problems are, what technical solutions are appropriate, and most difficult of all, how to make those solutions socially acceptable. To our frustration, the public seems to have lost confidence in the ability of politicians and professionals to solve the problems.

This chapter traces the history of the modern environmental movement through the rise and fall of types of expertise brought to bear on the problems. Then the relevance of the technical solution to society's means for addressing the problems is examined. Last, the formal methods established to make environmental decisions are examined and contrasted with the ways in which decisions are actually reached. This approach will lead to recommendations on how to proceed in the future.

Let us look first at the professions that have been in the forefront of environmental management over the years. Each has discovered problems and offered solutions to them. Unfortunately, some of these solutions have had unforeseen consequences. Almost all have been focused so narrowly that opportunities to do the job right have been lost. I have chosen illustrations from water resource management because that is the environmental problem most familiar to me. However, similar examples exist in air pollution and solid waste management.

Physicians were first to understand the direct effects of man's activities on water supplies. Their initial efforts were aimed at keeping harmful exposure to a minimum. Sewage was carted away from populated areas, and by 1850 the storm drain was commonly used to dispose of household wastes, thereby drastically reducing the spread of cholera. This solution led to the unforeseen consequence of dumping raw sewage into water bodies used as sources of drinking water. The response to the resulting threat to public health was chlorination of public water supplies to prevent diseases caused by this method of sewage removal. By 1930, chlorination had virtually eliminated typhoid in the United States. In a sense, this major achievement completed the contribution of the early public health approach to wastewater management. What have the consequences been? First, the large quantities of sewage that are diluted with water and shunted off to bodies of water cause eutrophication and contamination of our natural surface water bodies. Tremendous quantities of water and nutrients are wasted in this practice. Ironically, it has since been discovered that chlorine itself may be an indirect cause of illness by combining with organics to form trihalomethanes, which are suspected carcinogens. This problem has required attention at 3,000 public drinking water systems around the country.

Despite what appeared to be permanent public commitment to the discharge of domestic waste mixed with vast quantities of water, the engineering profession rallied to the task of reducing the nutrients and pathogens entering waterways. By 1970 a vast infrastructure of secondary treatment plants was substantially complete all over the country. Building continues on these facilities. Between 1977 and the present, local and federal governments have invested more than $100 billion to gain 87 percent compliance with the standards for secondary treatment. However, by 1980 policymakers and regulators realized that they had been lulled into a false sense of accomplishment. Despite this enormous infrastructure investment, few improvements have occurred in most of our waterways since those achieved by the early 1970s. It is not surprising that the public has lost confidence. As some engineers had warned, slipping by the neatly devised and heroically built system were 37,000 inappropriately designed landfills, hundreds of thousands of leaking gasoline tanks, and millions of tons of untreated nutrients and metals left over from secondary treatment. Worst of all, discharges of storm water, as polluted as raw sewage and laden with heavy metals and exotic chemicals, continue to run untreated into our lakes, rivers, and estuaries. Bizarre new fish diseases are appearing almost weekly around the country, most likely because of long-term bioaccumulation of unregulated pollutants. In Florida, such storm water accounts for all solids,

80–95 percent of the heavy metal loads, and 20 percent of the nutrients polluting our surface waters.

Despite the quantities of water treated and wasted in secondary treatment plants, new water supplies are continuously being sought. The U.S. Army Corps of Engineers has been active in this area in Florida, and many benefits have resulted. However, there are some sad legacies of the past. One is the Central and South Florida Flood Control project designed to create and protect 750,000 acres of agricultural land, formerly part of the great Everglades, and to provide water in times of drought to the urban areas of south Florida. Terrible by-products of this project have been a 90 percent reduction in the population of wading birds in Everglades National Park and the eutrophication of Lake Okeechobee, the heart of the freshwater supply to south Florida.

For their part, the biologists are waiting in the wings to solve these problems. If more had been known, they say, few of these disasters would have occurred. Absent a clearer understanding of consequences and alternatives, the public is now more respectful of embarking on new projects. However, after 20 years of study on Lake Okeechobee, biologists still cannot describe with any certainty the nutrient regime of the lake. They are similarly confused on issues surrounding the effects of acid rain and other major ecological disturbances. Few in our society believe it would be prudent to wait to intervene in such problems while biologists fully sort out causes and effects.

Because causes, effects, and cures are still elusive in many large environmental problems and enormous challenges keep appearing, the condition that has developed is obviously one in which the legal profession can flourish. The legal system has produced some of the basic decisions supporting environmental protection, but it has also produced an adversarial, combative climate in which it is impossible for people from industry to feel comfortable discussing facts with their colleagues in government or with the public. Many people who are knowledgeable about environmental issues are constantly in litigation and constrained from solving problems by using each other's talents cooperatively. The amount of litigation is alarming. For those cases that went to trial in federal court, 10 percent of the civil suits overall took longer than 45 months to resolve, and 10 percent of the environmental cases took longer than 67 months to resolve. Most serious is the fact that the legalistic approach has produced a staggering load of regulations, purportedly to cover every conceivable circumstance. This regulatory burden has left little time or incentive for creativity and human judgment, and no time for concentrating on environmental results. It has created a process-oriented, rather than a results-oriented, approach to environmental regulations.

The purpose of this review is to underscore the need for humility in proposing universal solutions to individual problems. Regulators have to consult with their colleagues and force themselves to justify carefully the need for action and its probable consequences.

## OPPORTUNITIES AND OBSTACLES

What lessons can engineers learn from these experiences? Engineers will continue to be hampered by a poor understanding of the biological world as reflected in the poor models of it. Research is essential to improve these models. Nevertheless, the engineering profession can move out with confidence and self-respect in developing several technological opportunities.

First, because experts are weakest in convincing each other, let alone the public, that they can describe and quantify the effects of contaminants released into the environment, there is a need for chemists and engineers to work diligently at finding new processes to avoid creation of these by-products. It is no longer possible to make radical improvements in end-of-process treatment. In large measure, the concept of treatment should become passé. Let us not give the biologists and the lawyers anything to worry about.

Second, recycling technologies will be required for unavoidable by-products and for reusing or reformatting products that are no longer useful. America generates two to three times more garbage than our economic peers do, and one-third of current landfills in the United States will be out of space in 5 to 10 years. New landfills and incinerators are becoming impossible to site. The only silver lining to this situation is that disposal costs have tripled or quadrupled in many locations, making recycling more palatable. The time has come to avoid product or process technologies that create new waste disposal problems in favor of those that reduce the need for ultimate discharge or disposal (see Friedlander, this volume). The developed world will not be allowed much longer to dump used articles on a poorer country in the name of recycling, if the result is to contaminate the land in that country.

Third, there is a need to plan for problems that are coming and effects that are unavoidable. It is not necessary to wait for the modelers to describe these effects in detail. The engineering profession can help find ways to reduce the carbon dioxide burden in the atmosphere where possible, but also to prepare for rising sea levels. Local coastal effects and the best mitigation for these impacts must be understood before disaster occurs.

Fourth, appropriate development of water supplies and efficient use of available resources are going to be of major importance as water becomes increasingly scarce in many parts of the world. It has been estimated that global warming, with an increase of 2°C in temperature and 10 percent reduction in rainfall, could reduce available water supplies by 50 percent in the drier states of the U.S. Southwest.

Finally, although many environments have damaged in the past, there is still time to rehabilitate many of them, including the Everglades and the Chesapeake Bay. It will take our best team efforts to bring these wonderful places back, yet we must because we depend on them economically and culturally and because they are natural wonders. There will be a host of new public works actions needed to correct non-point-source problems and delicate freshwater–saltwater imbalances and to restore the natural hydroperiod essential to a balanced fishery.

There are exciting and challenging opportunities for engineers in conserving natural resources, but it is equally important to ensure that these solutions are usable and used. Let us examine for a moment the unreceptive atmosphere in which these brilliant and practical new discoveries will struggle to live.

First, problem definition is often a major drawback to progress. There are usually deficient or conflicting scientific data defining the problem. The public often disagrees about the causes of problems and the priorities for solutions. This is not surprising because scientists also often disagree, both on sources and fates of contaminants in the environment, and on political aspects of the issues as well.

Second, appropriate solutions may elude us because the regulatory system and the market often do not encourage them. For example, we require advanced waste treatment of domestic waste at about 50 percent higher cost than the usual secondary treatment when discharged into a eutrophic water body. Right next to this "gold-plated pipe" is a storm water ditch carrying the equivalent of raw sewage. This water has received absolutely no treatment.

Third, some of the toughest environmental issues—ozone pollution from automobiles, climate change, eutrophication of water bodies, and loss of natural habitat due to the growing market for vacation homes—are the consequences of large-scale cultural patterns, the summed effects of millions of people making individual decisions. It is easy for people to rally around a common enemy "the smokestack," but ask them to separate their garbage or stop fertilizing their lawn and the commitment to environmental quality becomes less important.

Fourth, there is a common tendency to rely on "high-tech" solutions and use "low-tech" human beings to implement them. Three Mile Island, Bhopal, and Chernobyl all come to mind. This problem has arisen so often that a whole new discipline, "human engineering," has developed to cope with it. Let us not forget that even with a computerized cockpit, experienced pilots still forget to set flaps. Human frailties are here to stay, and design must involve an understanding of human behavior.

Fifth, the regulatory system was largely designed around the outmoded concept of treatment after process completion, rather than avoidance and reuse. This has been a more practical approach for the regulator and keeps government out of the internal workings of the regulated community. It does not encourage the modern approach, which by its nature is highly individualized by location, so that each plan is suited to the individual sensitivites of each natural system.

Last, despite a centralized approach to pollution control, one that is highly structured and legalistic, the United States is moving more and more toward negotiated decision making. Still another new profession, the environmental mediator, has leaped into the fray.

## THE REAL WORLD OF DECISIONS

Many times I have heard competent industrial managers say that they are frustrated by expensive regulations which they feel are irrelevant or by local citizens who fail to distinguish between a real risk and what is merely a fear. These managers feel betrayed by local and regional governments that feel compelled to add their own burden of regulation because, somehow, the state and federal governments are not doing their job. Local regulations often conflict with the national approach. This situation is a natural consequence of our lack of hard data and the mistaken demand on all sides for clearly articulated rules so that all parties can tell what is expected of the regulated party. The sheer volume and conflict among all the rules make that clarity a chimera.

As a result of this complex regulatory structure and the public's continued distrust, many of the real decisions are actually being made locally with a far broader agenda than that normally encompassed by the regulatory approach. Some people call this the "let's make a deal" approach. State and local authorities need to take advantage of this approach to encourage regional solutions to environmental problems, solutions tailored to the environmental needs of each area and to the causes of those problems. The Washington establishment will not always like this devolution of authority.

Although there are many examples of this regulatory approach around the country, I will discuss two that I have participated in. The first concerns a phosphate mining and chemical plant operated by Occidental Chemical Corporation at White Springs, Florida. The company owns mineral rights along the Suwannee River and for several miles inland. In 1984, new management at the plant became frustrated by the constant litigation surrounding every change contemplated. Planning expansion became impossible because of regulatory uncertainties and mounting public concern. The company decided to open a dialogue with concerned agencies and environmental groups. Although not every issue has been settled, certain things have become clear to everyone. First, the protection of the Suwannee River depends absolutely on developing a long-range plan that includes preservation elements, acquisition elements, and mitigation of damage to wetlands that are to be mined. The need for a long-range environmental plan coincided with the company's view that, to make long-range business plans, they had to know which areas would be allowed to be mined. Many of the items that are central to the agreement lie outside the normal scope of regulation, yet are essential to the well-being of the Suwannee River watershed. Of course, this is a simple example compared with areas that have more than one source of pollution.

A second example is the comprehensive basin approach to managing water problems, which was begun in Florida in 1986. In 1987 Florida passed the Surface Water Improvement and Management Act, which designated critical basins. The new aspects of this program are funded at $15 million per year, combined with $20 million per year for land acquisition programs. By contrast, the entire Clean Lakes program (authorized under the Federal Water Pollution Control Act Amendments of 1972, Public Law 92-500) in the U.S. Environmental Protection Agency receives only $15 million. On the St. Johns River near Jacksonville, for example, marsh restoration, land acquisition, water supply, flood control, and enforcement actions are being combined in a massive effort. It has become clear to us in Florida that it is impossible to meet environmental goals on a routine permit-by-permit basis.

We can listen carefully to these issues and examples and fashion a newer, better means of dealing with many environmental problems. It must be one that has a sound scientific base, has incentives for doing the right thing, engages people's cooperation early in the process, recognizes that humans are mortals, is relatively site specific and results oriented, and is negotiated and agreed to by all parties.

Engineers must now consider such interaction with the agencies and public as part of their job: the public should be the ultimate client for every environmental engineer. It is time again for engineers, as well as representatives of the many other professions with relevant expertise, to step forward and commit themselves to maintaining and enhancing environmental quality.

## BIBLIOGRAPHY

American Water Works Association. 1981. Water Conservation Management. Washington, D.C.: American Waterworks Association.
Bingham, G. 1986. Resolving Environmental Disputes. Washington, D.C.: The Conservation Foundation.
Costanza, R. 1987. Social traps and environmental policy. BioScience 37(6):407–412.

Elkington, J., and J. Shopley. 1988. The Shrinking Planet: U.S. Information Technology and Sustainable Development. Washington, D.C.: World Resources Institute.
King, J. 1985. Troubled Water. Emmaus, Pa.: Rodale Press.
Morgan, A. E. 1971. Dams and Other Disasters. Boston: Porter Sargent Publishers.
National Academy of Engineering. 1986. Hazards: Technology and Fairness. Washington, D.C.: National Academy Press.
National Academy of Engineering. 1988. Cities and Their Vital Systems. Infrastructure Past, Present, and Future. Washington, D.C.: National Academy Press.
National Council on Public Works Improvement. 1988. The state of U.S. infrastructure. Urban Land May:20–23.
Rabe, B. G. 1986. Fragmentation and Integration in State Environmental Management. Washington, D.C.: The Conservation Foundation.
Stokey, E., and R. Zeckhauser. 1978. A Primer for Policy Analysis. New York: W. W. Norton.

# Environmental Issues: Implications for Engineering Design and Education

SHELDON K. FRIEDLANDER

Pollution control in the United States has depended on regulatory action to drive changes in technology. The approach has been to trace the effects of pollutants on receptors (humans, ecological systems, etc.) back to the appropriate pollution sources and then implement control measures. The role of the technological community has been largely reactive.

The use of regulatory action to force technological change is often costly and inefficient. It starts on the outermost perimeter of technology where the effects are detected and it encourages end-of-pipe treatment. It is ad hoc, often limited to special classes of pollutants, and leads to a distorted view of industry as a collection of pollution sources.[1]

Despite some successes with the effects-based approach, environmental problems of technological origin and an evermore critical public attitude toward technology continue. Regulatory trends indicate no letup in restrictions on the environmental side effects of technology. However, both the regulatory agencies and industry appear receptive to a more proactive mode of operation by industry. This calls for the accelerated development of systematic procedures for the design of environmentally compatible technologies. For this purpose, it is necessary to go back to basics and make institutional changes affecting engineering design, research, and education as discussed in this chapter.

## REGULATORY TRENDS

In the short term, the requirement for industry to comply with federal right-to-know laws by disclosing emissions of toxic chemicals will stimulate more stringent state and local laws (*Wall Street Journal*, July 28, 1988). Pressures on industry to reduce emissions will mount as the levels required for disclosure decrease.

For the long term, there is a growing movement toward regulation at an international level: as long as the effects of pollution seemed limited to a scale of a few miles in the vicinity of a source, regulation was the responsibility of local and state governments. Later, overall regulatory responsibility in the United States was assumed by the federal government through the Environmental Protection Agency.

A new precedent for regulatory action has been set at the international level by the Montreal Protocol on Substances That Deplete the Ozone Layer (Doniger, 1988; United Nations Environment Program, 1987). The commercial use of these compounds began in the early 1930s. The common refrigerants at the time—ammonia, methyl chloride, and sulfur dioxide—were not suitable for home refrigeration because of their noxious or toxic properties. In the late 1920s, Charles F. Kettering of General Motors asked one of his research staff, Thomas Midgley, Jr., to find a nontoxic, nonflammable substitute for these gases (Kettering, 1947). Midgley, educated as a mechanical engineer at Cornell University, was self-taught in chemistry. In 1930 he prepared dichlorofluoromethane (Freon 12) and demonstrated its safety at a meeting of the American Chemical Society by inhaling a deep lungful of the gas and then exhaling it to extinguish a lighted candle. Midgley's discovery laid the foundation for the success of the Frigidaire Division of General Motors. During World War II, freons were used to aerosolize insecticides such as DDT. Later they found widespread use as solvents in the microelectronics industry. Theoretical studies by Mario J. Molina and F. Sherwood Rowland in the early 1970s, subsequently confirmed by large-scale atmospheric observations in the 1980s, indicated that damage to the stratospheric ozone layer resulted from these long-lived substances (Molina and Rowland, 1974).

The Montreal Protocol requires signatory nations to freeze production of five CFCs at 1986 levels and then cut production in half by July 1, 1998 (see Glas, this volume). Provision is made for the state of technological development of the participating nations. Developing nations are allowed a 10-year growth period for CFC and halon use, and producing countries can make extra CFCs and halons, not counted in their national quotas, for the developing nations. The Soviet Union also receives a catch-up allowance, similar to the developing nations.

This accord sets two important precedents: it recognizes the atmosphere as a limited, shared resource; and it curtails the right of individual countries to release wastes to the atmosphere. In effect we have agreed to the allocation of rights to release atmospheric emissions on a national basis for CFCs.

If the Montreal Protocol is used as a model for controlling other types of pollutants with regional and global effects, harder decisions must be faced. It is much more difficult for a modern industrial society to curtail releases of carbon dioxide and the oxides of nitrogen ($NO_x$), due to burning of fossil fuel than to control CFCs.

Currently, the United States and Europe are the major sources of $NO_x$ emissions, followed by East Asia (Figure 1). The developing countries in southerly latitudes, however, as well as East Asia, had the largest percentage increases in $NO_x$ emissions from 1966 to 1980 (Figure 2). These increases portend a substantial spread in $NO_x$-affected areas around the world, with accompanying photochemical smog and acid deposition.

Developing nations have a much larger population base and much faster-growing populations than the United States and Europe (Figure 3). As these nations industrialize, worldwide emissions of all pollutants will grow rapidly with the following consequences:

- pressure for further regulation at the international level;
- difficult negotiations between industrialized nations and developing countries in setting national emissions allocations; and
- a competitive advantage to nations whose engineers are able to design clean, economic technologies.

Environmental protection on a global scale will require the industrialized nations to transfer low-pollution technologies to developing countries in a timely manner.

## A CHALLENGE TO ENGINEERING

Before discovering Freon, Thomas Midgley had also worked on the problem of engine knock at Kettering's request. In 1921 Midgley and his research team discovered tetraethyl lead. This technological breakthrough led to high-octane gasoline, which permitted the use of more fuel-efficient, high-compression engines. Midgley went on to develop Freon, was elected to the National Academy of Sciences, and later became president of the American Chemical Society.

Midgley did not live to see the controversies surrounding the environmental effects of his great discoveries. In 1940 he suffered an acute attack of poliomyelitis, which left him crippled. He set up a pulley and harness system to assist him into and out of bed, but tragically strangled himself in the harness in 1944. Midgley's distinguished career, important technological contributions, and tragic death read like a parable—a moral tale about how our marvelous technology designed to satisfy the needs of society may unleash unanticipated harmful effects on the same society.

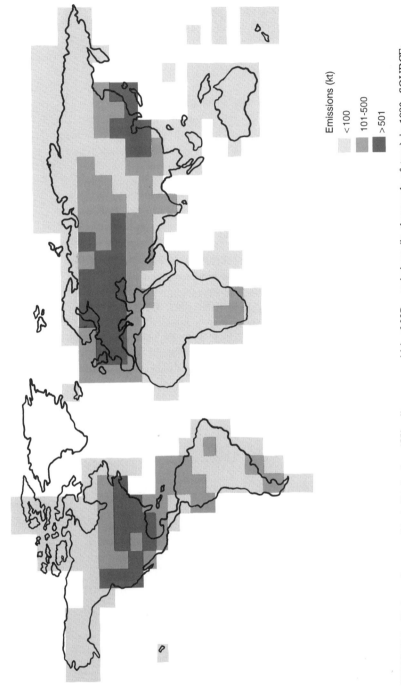

FIGURE 1 Estimated geographical distribution (550-mile-square grids) of $NO_x$ emissions (in thousands of tons) in 1980. SOURCE: After Hameed and Dignon (1988).

The engineering profession faces the challenge of satisfying those societal needs while meeting increasingly strict regulations on environmental side effects. Over the long term, this calls for institutional changes affecting engineering design, basic research, and education. Recent efforts to reshape technological approaches to pollution control (National Research Council, 1985; U.S. Office of Technology Assessment, 1986) have focused on process and plant design, which are considered first in this chapter. Consumer products, from automobiles to plastic wrappings, have widespread environmental effects and require separate consideration. Finally, the implications of these technological challenges for engineering education are discussed.

Structural changes in technology will have profound effects in the context of this discussion. For example, in the chemical industry, there is an anticipated shift from commodity chemicals to high-value, low-volume specialty chemicals. In the field of energy, superconductivity and fusion, now at early stages of development, appear to be clean technologies. As new technologies become familiar, however, nasty surprises tend to occur. Moreover, if new energy technologies are not viable scientifically or economically, and another energy crunch develops, waiting in the wings are coal-derived fuels and nuclear energy, each of which has its own environmental consequences.

Efforts to predict structural changes in technology (see Ausubel, this volume) should continue. In my view, however, it is unlikely that we will be able to predict in much detail, very far in advance, the onset of new technologies and their environmental effects. Over the long term, however, institutional changes affecting engineering design, basic research, and education are necessary to shape the development of new technologies with an understanding of potential environmental consequences.

## DESIGN OF ENVIRONMENTALLY COMPATIBLE PROCESSES AND PLANTS

A recent National Research Council report (1988, p. 112) summarizes the challenges faced in the design of environmentally compatible manufacturing plants as follows:

> Traditional analyses of process economics might show that inherently safer and less polluting plants are less efficient in terms of energy or raw materials usage. Indeed, chemical plants have been designed in the past principally to maximize reliability, product quality, and profitability. Such issues as chronic emissions, waste disposal, and process safety have often been treated as secondary factors. It has become clear, however, that these considerations are as important as the others and must be addressed during the earliest design stages of the plant. This is in part due to a more realistic calculation of the economics of building and operating a plant. When potential savings from reduced accident frequency, avoidance of generating hazardous waste that must be disposed of, and decreased potential liability are taken into consideration, inherently safer and less polluting

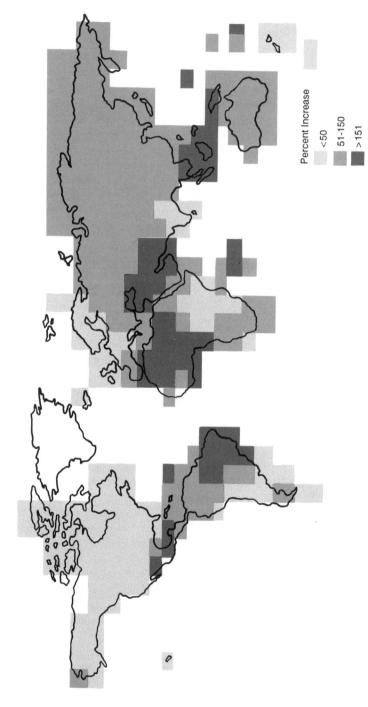

FIGURE 2 Percentage increases in $NO_x$ emissions from 1966 to 1980 in the developing countries in the central latitudes as well as in East Asia. SOURCE: After Hameed and Dignon (1988).

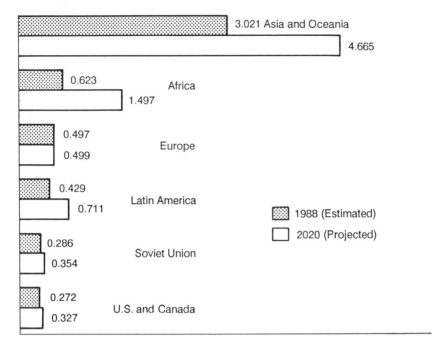

FIGURE 3  World population growth (in billions). SOURCE: Population Reference Bureau, Inc.

plants may prove to cost less overall to build and operate. And in any case, if the American public is not convinced that chemical plants are designed to be safe and environmentally benign, then the fact that they operate economically will be of little consequence to the public's decision on whether to allow their construction and operation.

Although chemical plants are singled out, this discussion is applicable to any type of manufacturing plant.

In the past, the design of environmentally compatible manufacturing plants has generally meant the use of end-of-pipe treatment or separation devices through which effluent gases or liquids pass on their way to the environment. These devices are designed to meet government emission standards for particular chemical compounds. However, accompanying nonregulated substances almost always remain in these streams. Ash and other solid wastes must be disposed of separately.

Over the past few years, a movement has grown stressing in-plant practices (as opposed to add-on devices or exterior recycling) to reduce or eliminate waste. This movement has been called waste reduction (National Research Council, 1985; U.S. Office of Technology Assessment, 1986) or production-integrated pollution control.

This approach is not new. For example, in a pioneering air pollution study, Rupp (1956) wrote:

> Source control and abatement of formed contaminants are complementary practices in the campaign against air pollution. Source control prevents the emission of contaminants to the atmosphere; abatement renders the emission of contaminants to the atmosphere harmless and inoffensive.

Waste reduction can be defined (U.S. Office of Technology Assessment, 1986) as "in-plant processes that reduce, avoid and eliminate" the generation of waste. Actions taken away from the manufacturing activity, including out-of-plant waste recycling, or treatment and disposal after the wastes are generated, are not considered waste reduction in this formulation, nor is concentrating wastes to reduce their volume. Opinions differ on the exact definition of waste reduction, but the general approach is clear.

The case for primacy of waste reduction rests on several factors: avoiding formation of a waste eliminates the need for treatment and disposal, both of which carry environmental risk. Control technologies may fail or fluctuate in efficiency. Treated effluent streams carry nonregulated residual substances that may turn out to be harmful. Secured disposal sites eventually discharge to the environment.

Five methods of waste reduction were identified in a study by the U.S. Office of Technology Assessment (1986, p. 27). They are listed here in order of decreasing use as reported in an industry survey:

1. in-plant recycling;
2. changes in process technology;
3. changes in plant operation (e.g., suppression of fugitive emissions);
4. substitution of input (raw) materials; and
5. modification of end products to permit the use of less polluting upstream processes.

Although waste reduction is an attractive concept, the total elimination of manufacturing wastes is beyond the capability of modern technology. The issue is really how to approach this limiting goal in an expeditious and cost-effective manner. For this purpose, regulatory action will undoubtedly play a role; it is in the national interest for the engineering community to guide such regulation to be sure that it is as rational as possible and cost-effective. From a technical point of view, research and development are of special concern.

## RESEARCH IMPLICATIONS

Treatment and disposal technologies are generally categorized according to the scientific or engineering principles on which they are based.

For example, chemical destruction methods are based commonly on combustion and biochemical (microbial) processes. Separation technologies employ filtration, electrical precipitation, scrubbing, and other recognized physicochemical processes to collect or concentrate wastes before destruction or disposal. Although many technical problems remain, a generally accepted framework exists for guiding research and development to improve the performance of waste destruction and separation technologies.

In contrast, the technology of waste reduction as an alternative to treatment and disposal does not have a widely accepted scientific basis. Proponents of this approach (National Research Council, 1985; U.S. Office of Technology Assessment, 1986) have made good use of case studies to illustrate the concept and its application to engineering practice. The challenge in the next phase will be to develop guidelines for basic engineering research underlying waste reduction.

As director of the Engineering Research Center for Hazardous Substance Control at the University of California, Los Angeles, I recently consulted with members of our industry-government advisory board on research in waste reduction. Several board members expressed pessimism about academic engineering research in this field. They argued that waste reduction is process specific and that the proprietary nature of the processes makes it too difficult for academic engineers to obtain enough information to make a contribution. Other members of the advisory board believed that increasing regulatory pressures were likely to force disclosures of information on such processes anyway.

In my view it is quite possible—in fact, essential—to do basic engineering research in support of waste reduction activities. Indeed such research has been done in combustion science for many years. A case in point is the control of $NO_x$: basic research on $NO_x$ formation in combustion has been driven by Arie J. Haagen-Smit's discovery of the key role that $NO_x$ plays in the generation of photochemical smog (Haagen-Smit, 1952) and by the importance of nitric acid in acid deposition. We now know that in furnaces and boilers, $NO_x$ emissions come mostly from fuel-bound nitrogen compounds; $NO_x$ emissions from internal combustion engines come mainly from the oxidation of atmospheric nitrogen. Better information on the fundamentals of the combustion process has made it possible to design for lower $NO_x$ emissions.

Another advance in basic engineering understanding is the dual-mechanism model for particle formation in coal combustion (Figure 4), which Richard C. Flagan and I developed quantitatively (Flagan and Friedlander, 1978). Submicron particles form by coagulation of nuclei produced by condensation of a small volatile fraction of the ash; coarse particles result from ash inclusions in the pulverized coal. This model, now well

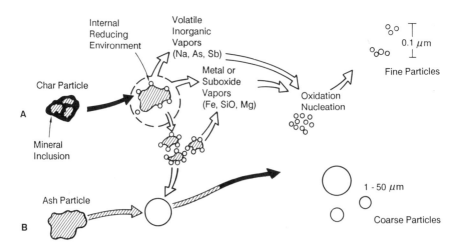

FIGURE 4 Dual-mode mechanism for formation of both fine and coarse particles in coal combustion. (A) fine particles enriched in potentially toxic metals form by condensation of a small amount of volatilized ash; (B) coarse particles result from ash inclusion in the pulverized coal. Analogous processes probably occur in the incineration of municipal and hazardous wastes. SOURCE: Flagan and Friedlander (1978).

established experimentally, offers new possibilities for reducing particulate emissions by modifying the particle size of pulverized coal and the combustion conditions.

A class of generic scientific or engineering principles must be developed, as in the combustion science examples given above, on which a fundamental engineering approach to waste reduction can be based. For this purpose, the following examples can be given:

- Chemical manufacturing processes may employ alternative chemical reaction paths and raw materials to reach the desired reaction products. The use of alternate reaction paths as a generic approach to waste reduction has recently been discussed in *Frontiers in Chemical Engineering* (National Research Council, 1988, p. 112):

    A chemical synthesis tree graph with a high-value product at its apex, lower-value raw materials at the base, and reaction steps as nodes connecting all branches offers a basis for quantitative assessment of feasible and economic process alternatives. It could also serve to define the safety and environmental impact of a pathway and offer a basis for safe designs that produce minimal wastes.

    This approach offers the possibility of selecting the chemical system and operating conditions (temperature and pressure) for minimum production of harmful by-products.

- The design of in-plant recycling systems optimized to minimize effluent streams is another area of basic engineering research important to waste reduction. At the start it will be desirable to formulate the problem in general terms, perhaps by analyzing networks of separation units analogous to the heat exchanger network approach developed to minimize industrial energy consumption.
- General principles must be sought to guide the search for substitutes for certain broad classes of widely used materials with potential environmental effects. For example, solvents and cleaning agents are used in almost every branch of industry. For the near term, there is a need for substitute solvents to replace halogenated organic compounds in many applications. This prospect provides an incentive for the synthesis of new types of solvents and for fundamental studies of interfacial phenomena involved in solvent action.

These few examples are illustrative. By identifying and supporting a broad set of basic research areas underlying waste reduction, more of our best research engineers and scientists can be encouraged to participate in this challenging task. Encouraging university research of this type will ensure that our students—the next generation of engineers and managers—better understand the issues involved.

## ENVIRONMENTALLY COMPATIBLE CONSUMER PRODUCTS

Some consumer products have major environmental consequences either during use (automobile or aerosol sprays) or after disposal (plastic containers, paints, and solvents). Indeed, there is growing evidence that the environmental consequences of consumer products may be more important than the direct effects of industrial activity:

- Consumer products are by their nature dispersed widely through the society, and they and their environmental effects remain in close contact with the population at risk.
- The average consumer has little technical know-how and cannot be expected to deal individually with complex chemical problems.
- Consumers as a group, when mobilized, have enough political power to limit efforts to control their own pollution-producing activities.

A case in point is plastic products, including containers and wrappings made from polymeric materials. For durability and low toxicity, these products are designed to be less reactive chemically. Plastics now constitute about 7 percent by weight of all municipal waste, and this figure is expected to rise to 10 percent by 2000 (Crawford, 1988). Plastic products accumulate in municipal landfills, on beaches, indeed throughout the environment. Communities in California and New York have already begun to ban certain types of plastic packaging, and about a dozen states are considering restrictions on their use (*Wall Street Journal*, July 21, 1988).

In response to developing regulatory trends and competition from the paper industry, the chemical industry in the United States and Europe has begun the development of biodegradable plastics much as was done 25 years ago when long-lasting detergents were polluting water supplies. One approach to biodegradable plastics incorporates cornstarch as an oxidizing agent in the polymeric materials. Oxidizing agents react in the presence of metal salts in the soil to degrade the polymers. The plastic becomes porous and brittle in 2 years; and after 10 years, the fragments are too small to see with the naked eye.

Degradable plastics carry with them their own risk. Their life may be limited. The degradation products may have harmful effects. The integrity of such products as plastic lumber or piping, which are made from recycled plastics, will be jeopardized if both degradable and nondegradable plastics are mixed during recycling.

The market for recycled plastics is currently strong (Crawford, 1988). The price of virgin resin has increased because U.S. production facilities are operating at capacity. As an example, the recycling of polyethylene terephthalate soft drink bottles has grown from 8 million pounds in 1979 to 130 million pounds in 1986. The potential revenue from plastic recycling is estimated to equal that from the recycling of newspapers, about $300 million annually.

To what extent will the public participate in the separation and collection steps required to recycle plastics? Few states expect that more than half of their plastic wastes can be recovered in this way. New Jersey has set a goal of 25 percent recovery; the remainder will continue to be incinerated or sent to landfills.

The incineration of plastics has two significant disadvantages. First there is the destruction of a reclaimable resource. Second is the possible emission of hazardous air pollutants from incinerators burning chlorine-containing plastics, especially polyvinyl chloride.

Thus, plastic consumer products are a part of a complex environmental problem that includes the consumer, the manufacturer, and the government. The relative importance of the environmental effects of consumer products compared with the direct consequences of industrial activity need to be evaluated systematically to help set environmental priorities (see Ayres, this volume). The ability of industry to design environmentally compatible products will strongly affect the extent of government intrusion into the marketing of such products.

## IMPLICATIONS FOR ENGINEERING EDUCATION

Not long ago the head of the hazardous substance control program at a large company met with the engineers of one of the production units to explore possibilities for process-integrated pollution control. The production engineers did not take easily to the idea of incorporating pollution control into the original process design. This may be an isolated incident, but my guess is that it is a fairly widespread attitude in industry, one that goes back to the early educational experience of engineers. This underlines the need to incorporate these concepts early, starting with the undergraduate engineering curriculum. How can this best be accomplished?

I see no need for a new branch of engineering to deal with the design of environmentally compatible technologies. Indeed, I believe that would be counterproductive for the following reasons. The selection and design of manufacturing processes and products should incorporate environmental constraints from the start, along with thermodynamic and economic factors. This is best done by the chemical, civil, mechanical, and electrical engineers, and others charged with process and product design in their respective industries.

Similarly, concepts related to pollution control should be incorporated into the normal curricula of the separate engineering disciplines. For example, concepts related to $NO_x$ emissions can easily be introduced into discussions of power cycles in undergraduate engineering thermodynamics classes.

As a part of an education in chemical engineering, courses in separation processes, chemical reaction engineering, and especially, the senior-level design course should all incorporate problems and examples related to minimization of pollution. There should also be a place for a few separate courses in pollution control.

At the University of California, Los Angeles, there is a separate chemical engineering undergraduate elective course in pollution control, and an option in environmental chemical engineering is being developed as part of our fully accredited chemical engineering undergraduate program. This course includes not only process and plant design, but also environmental transport and transformation, public health, and ecological effects.

As engineers, we naturally focus on technological systems, which are our particular responsibility. At the same time, engineers should have a good understanding of the interaction between the technological and environmental systems. We must be able to meet specialists in public health and ecology halfway to put together as complete a picture as possible of the environmental effects of technological systems. We have a major role to play in relating environmental quality to industrial emissions through receptor and dispersion models. Engineers should have a good understanding of the ecological or health basis of the regulatory standards that apply to the systems they are designing.

Students tend to be idealistic and are attracted to the idea of developing environmentally compatible technology, but engineering students are also attentive to the signals sent by industry. Industry has a stake in encouraging students to acquire the skills in engineering design needed to protect the environment. Industry should make clear its commitment to the systematic incorporation of pollution control in engineering education through the professional societies and university curriculum accreditation procedures.

## SUMMARY

The engineering profession faces a challenge—to satisfy societal needs with ever-tightening regulation of environmental side effects. This calls for new approaches in engineering education, basic research, and design. Engineering education should routinely incorporate environmental constraints into the design procedures of existing engineering disciplines, and environmental consequences of technology and the basis of regulatory standards should be part of the engineering curriculum. In plant design, there must be a growing emphasis on waste reduction rather than end-of-pipe treatment and disposal. However, waste reduction needs a fundamental engineering research base, which is still under development and merits high priority. Finally, systematic improvements in consumer product design are required, especially for items widely dispersed through society.

## NOTE

1. This attitude has been expressed many times in multidisciplinary conferences and workshops involving public health specialists, ecologists, lawyers, economists, and regulators.

## REFERENCES

Crawford, M. 1988. There's (plastic) gold in them thar landfills. Science 241:411–412.
Doniger, D. D. 1988. Politics of the ozone layer. Issues in Science and Technology 4:86–92.
Flagan, R. C., and S. K. Friedlander. 1978. Particle formation in pulverized coal combustion—A review. Pp. 25–29 in Recent Developments in Aerosol Science, D. T. Shaw, ed. New York: John Wiley & Sons.
Haagen-Smit, A. J. 1952. Chemistry and physiology of Los Angeles smog. Industrial and Engineering Chemistry 44(6):1342.
Hameed, S., and J. Dignon. 1988. Changes in the geographical distributions of global emissions of $NO_x$ and $SO_x$ from fossil-fuel combustion between 1966 and 1980. Atmospheric Environment 22(3):441–449.
Kettering, C. F. 1947. Thomas Midgley, Jr., 1889–1944. Pp. 361–376 in Biographical Memoirs, Vol. 24. Washington, D.C.: National Academy of Sciences.
Molina, M. J., and F. S. Rowland. 1974. Stratospheric sink for chlorofluoromethanes: Chlorine atom catalyzed destruction of ozone. Nature 249:810–812.

National Research Council. 1985. Reducing Hazardous Waste Generation: An Evaluation and a Call for Action. Board on Environmental Studies and Toxicology. Washington, D.C.: National Academy Press.

National Research Council. 1988. Frontiers in Chemical Engineering: Research Needs and Opportunities. Board on Chemical Sciences and Technology. Washington, D.C.: National Academy Press.

Rupp, W. H. 1956. Air pollution sources and their control. Pp. 1–2 in Air Pollution Handbook, P. L. Magill, F. R. Holden, and L. Ackley, eds. New York: McGraw-Hill.

United Nations Environment Program. 1987. Montreal Protocol on Substances That Deplete the Ozone Layer. Final Act. September 16, 1987. [Reprinted as appendix to U.S. EPA Proposed Rules for Protection of Stratospheric Ozone, 52 Federal Register (December 14, 1987):47489–47523.]

U.S. Office of Technology Assessment. 1986. Serious Reduction of Hazardous Waste: For Pollution Prevention and Industrial Efficiency. Washington, D.C.: U.S. Government Printing Office.

Wall Street Journal. July 21, 1988. Back to the lab: Big chemical concerns hasten to develop biodegradable plastics. P. 1.

Wall Street Journal. July 28, 1988. The environment becomes big business as laws get tougher. P. 1.

# Engineering Our Way Out of Endless Environmental Crises

## WALTER R. LYNN

In our society, crises play a vital role in the never-ending game of capturing the attention of the public. During the summer of 1988 the news media helped to make everyone who reads or watches television aware of a variety of unpleasant, costly, and disturbing events, all of which reflect the continuing crises of long- and short-term environmental changes. All the environmental news sounded bad and seemed to promise to become worse: droughts, floods, forest fires, solid waste washing up on public beaches, sewage pollution of water supplies, ozone depletion, the "greenhouse effect," acid rain, and more.

This chapter argues that, as important as public awareness is to solving problems related to environmental change, public and private energies should be redirected from following crises to opening avenues for more constructive response. Future crises can be averted through timely responses to anticipated and precursor conditions. The most effective control technologies are likely to result from local and regional actions guided by national and international consensus. The leadership to develop these technologies must come from government and industry, supported by science. We as individuals have important roles to play because, ultimately, technology is socially constructed, and the interplay of our views will filter, moderate, and determine what is acceptable.

### ENVIRONMENTAL ENDS, TECHNOLOGICAL MEANS

In the spring of 1987 new names invaded the public consciousness—*Mobro, Khian Sea, Bark*. The *Mobro*, a modern day "Flying Dutchman," spent two months cruising the Gulf Coast and the Caribbean in search of a final resting place for its unsightly and fragrant cargo of solid waste from Long Island. Other, less well known vessels, such as the *Khian Sea* and the *Bark*, wandered the Caribbean and the African coast carrying incinerator ash from Philadelphia. After fruitless searches for places that would accept their unsavory cargoes, the vessels returned to their home ports—at least, temporarily.[1] Other developed countries also have sought less expensive, "simple" solutions to their disposal problems, such as exporting them to poorer countries. Such actions continue even though international groups such as the Organization of African Unity characterize the export of toxic wastes to their continent as "a crime against Africa and the African people."[2]

That hapless scow the *Mobro* triggered an awakening of the public to the long-standing solid waste disposal crisis posed by the condition and capacity of sanitary landfills in the United States (National Research Council, 1984). Additionally, this disposal option, chosen by most municipalities because of low cost and convenience, no longer appeared to be a viable solution for the future because domestic wastes were often found to contain hazardous materials. State and federal requirements for land disposal imposed more stringent and costly controls at these sites. Regrettably, alternative technologies for treatment and processing of solid wastes are not competitive in cost and convenience with former sanitary landfill design and operations.

Confronted with the prospect of closing filled or nonconforming landfills, more stringent federal and state design and operational requirements, and rebellious communities unwilling to tolerate the construction of new ones in their vicinity, public works officials sought other alternatives to landfills, such as loading garbage on trucks, trains, boats, or barges for shipping to some far-off place or shifting to newer incineration technologies.[3,4] Shipping solid wastes someplace else in the United States turned out to be expensive and was not greeted graciously or with much enthusiasm (*Public Works*, 1988):

> Whether it's solid waste from historic Philadelphia, classy garbage from New York City's Fifth Avenue, or trash from the finely landscaped lawns of Northern New Jersey, the fact is nobody else wants it.

The summer of 1988 was special in the annals of environmental history. The United States experienced relatively severe drought conditions in much of the Midwest and elsewhere, and the drought was characterized by some as evidence of a more significant impending crisis—the "greenhouse effect."[5] Presidential candidates helped ensure that all Americans became aware that garbage was washing up on East, West, and Gulf Coast beaches. Weekly news magazines presented the story of beachfront pollution while deftly interweaving other local, national, and global pollution problems such as the hole in the ozone layer, acid rain, and disposal of nuclear wastes. Opinion polls indicated a large majority of public opinion favored stronger actions to preserve or enhance environmental quality.

Echoing these themes in an editorial, Harold M. Evans (1988) of *U.S. News and World Report* took the National Research Council and its parent Academies to task for their alleged complacency in characterizing concerns about chlorofluorocarbons, carbon dioxide, and acid deposition as "unwarranted and unnecessarily alarmist." Evans concluded that the conservatism of National Research Council panels and committees has consistently "thwarted pollution controls that would have cost millions at the time, but now confront us with costs of untold billions for irreversible consequences that might yet produce global catastrophe." The middle ground on environmental issues appeared to shift toward environmental activism reminiscent of the early 1970s, with the new element that global concerns matched the traditional ones close to home.

The universe of environmental changes, for the purpose of discussion, can be divided into two classes: global and local. Local changes have the following characteristics:

- They are often obvious (they can be seen, smelled, felt, etc.).
- The factors that cause these changes are reasonably well understood.
- The means to improve environmental conditions and to prevent further environmental degradation are relatively well known.

Global environmental change results from the cumulative effects of countless individual and collective actions at the local level. When there are no perceptible local effects, many individuals assume that what they do has no global effects. Where effects are difficult to detect with one's senses, they must be understood in the abstract.

Polyethylene wrappers provide an example of how little attention individuals pay to waste flow problems when making decisions. Polyethylene is essentially nonbiodegradable when it is deposited in sanitary landfills, and it is a source of concern when burned in municipal incinerators. Relatively recently a new product appeared in the form of clear polyethylene magazine wrappers. This is a useful product. It ensures that publications delivered by the U.S. mail system arrive in good condition. It also ended a long-standing complaint by some of the readers of *Science* who were annoyed because the mailing label was pasted on what they claimed was the otherwise attractive and useful magazine cover.

Although most people were probably content with the delivery system that existed before this product was introduced, after the fact, the widespread acceptance of this product is evidence of a "consumer demand"

—an unstated, perhaps unknown and obviously unfulfilled need now being met. It is hard to believe that the producers, distributors, or ultimate consumers of this product gave any consideration to the waste processing and social costs associated with it. Presumably, the improved physical condition of our magazines comes at the cost of increased waste flows and waste treatment.

Engineering communities have become painfully aware that such phrases as the "tragedy of the commons" (Hardin, 1968) and the "tyranny of small decisions" (Kahn, 1966) are not only parabolic but also accurate descriptions of reality. The engineer's consciousness about environmental issues was raised in 1970 with passage of the National Environmental Policy Act, which, among other features, established the Environmental Impact Statement (EIS) process. The EIS imposed an obligation on all engineers to consider the short- and long-term environmental consequences of various projects. The EIS process—which now exists in various forms in many states and even at local levels—requires those who are advocates of change to evaluate, disclose, and minimize adverse environmental consequences. Those requirements have helped to make environmental consequences as much a part of the overall engineering design process as are considerations of safety, economics, and useful life. Today, no responsible engineer designs anything without giving explicit consideration to its possible impact on the environment.

Welfare economists have long tried to convince us that the true costs of environmental consequences are classic examples of external economic effects and, thus, do not directly influence or play a role in guiding the decisions members of a society routinely make—decisions that may have devastating environmental outcomes. Because market forces do not consistently provide the kinds of messages that lead to sensible environmental outcomes, regulation is employed by societies to redress these problems. If one of our objectives is to prevent deterioration of environmental conditions for reasons of health and safety, then regulations compelling all producers to include the costs of pollution control in the prices of all goods and services would have a salutary effect on the behavior of organizations and individuals. It is true that regulations reduce our freedom of choice, but so does a deteriorating environment.

Until recently, it has been difficult to collect compelling scientific evidence to demonstrate that global environmental changes are taking place, and despite the new information, not everyone is convinced. Although there is even less agreement about what could or ought to be done to prevent global environmental change or to reverse changes that have already occurred, there is little argument that these kinds of changes result from ubiquitous local or regional actions. It seems clear that the only way such global issues can be addressed is if they are properly orchestrated at the national and international levels—guiding and, if necessary, directing local and regional efforts.

If mankind has become endangered as a result of significant global environmental changes that have already occurred, whose effects due to past actions are largely irreversible (at least for the next few decades), it is as important to focus on actions that *prevent* conditions from getting worse in the future as it is to clean up existing conditions. It is difficult to get political bodies to address and resolve environmental issues in their own backyards. Getting them to make behavioral changes and economic sacrifices in order to come to grips with global issues presents an enormous challenge.

Clearly, engineers have a responsibility, if not a duty, to act in ways that help reduce the likelihood of such potentially harmful environmental events. Although engineers are probably thought of as consummate technological optimists, there are things they can and cannot do through technology. Engineers must be constantly aware, and ultimately must convince the public, that technology is a means—not an end.

Over the past two decades, we have learned a great deal about the limits of "technological fixes." Technological optimism frequently led us to exceed unintentionally our competence and wisdom. Alvin M. Weinberg (1966) argued that some "quick technological fixes" were viable alternatives to "social engineering." Acknowledging that such approaches "do not get to the heart of the problem," are at best "temporary expedients," and "create new problems as they solve old ones," Weinberg claimed that changing people's behavior (i.e., social engineering) was a far more difficult and demanding task. Thus, even temporary technological patches were highly desirable because they would buy time and accelerate evolutionary change.

Over the past two decades, society has benefited a great deal from its ability to devise technological fixes, whereas the social and political processes have accomplished relatively little by way of "social engineering." However, these short-run solutions have not brought us much closer to confronting successfully the major environmental changes with which we and future generations have to deal.

When Weinberg suggested that technological fixes buy time, he implied that more long-lasting solutions were identifiable (or knowable) but were inaccessible for a variety of reasons, including costs and lack of scientific understanding. Currently, technology provides the only viable means by which our complex, interdependent society is able to address these environmental problems. Until those of us who create and devise these methods are challenged to put much more effort into preventing future adverse consequences, much of the engineer's contributions will be perceived as ineffectual in addressing the root causes of environmental degradation.

Regrettably, technological fixes are prescribed primarily to keep existing systems working. Little attention is given to determining and developing longer-term solutions, and short-run fixes become the order of the day. The result, as expected, is to move from crisis to crisis. Long-run solutions do not arise from technology alone; thus, we must look for answers that arrange decent marriages between social engineering and technology (Gray, this volume). For such marriages to be successful, the following conditions must be met:

- We must overcome the reluctance to recognize the existence of these environmental problems.
- We must educate individuals about how their behavior in exercising consumer preferences affects local and global environments.
- We must be prepared to spend money to develop the knowledge base needed to expand our understanding of the environment and to develop technological and social means to address these environmental issues.

## CHALLENGES AND OPPORTUNITIES FOR WASTE TECHNOLOGY

Although technology *alone* cannot provide long-term solutions to the kinds of problems we face, there is much important work to do (U.S. Environmental Protection Agency, 1987). There are opportunities to make significant improvements in solid, gaseous, and liquid waste treatment processes that meet elevated performance requirements. One of the most difficult challenges is to devise methods of treatment and disposal that can cope with smaller and smaller concentrations of impurities in waste streams and to accomplish that end without breaking the bank. The relatively easy, inexpensive treatment or disposal methods have already been devised and exploited, and the problems that remain are much more difficult and costly to solve.

Research and development are being carried out in places where they have always been done: university engineering and science departments and centers, state and federal research laboratories, and in private industry. Given the importance placed on competitiveness and productivity, it is distressing that so little attention has been given to the R&D effort needed not only to enhance our "environmental condition," but also to provide the science and technology required to support innovative production technologies and practices, many of which bring with them new hazardous waste problems.

The Engineering Research Board of the National Research Council (1987, p. 142) called attention to the need for the federal research support system to recognize that environmental resources are critical to the domestic economy, to national security, and to both human welfare and the quality of life in the United States. These resources are fundamental to other technologies as both inputs . . . and output. . . . As such, they form the base on which virtually all other economic activities are built.

Although statutorily mandated responsibilities have grown, support for R&D in these areas (identified in the federal budget as resources and environment) has declined to a level that makes one apprehensive about the capacity of the engineering and scientific research communities to sustain a meaningful research agenda to address these problems.[6] The record clearly shows that federal support for research in the areas of resources and environment has greatly diminished in the 1980s. The time has come to develop an R&D program that truly represents a national commitment to address the threat, if not the clear and present danger, posed by environmental changes at both local and global levels. Without such support it will not be possible to provide the technological base required to cope with the ever-changing and expanding demands to which society must be prepared to respond.

A broad range of R&D topics must be explored to gain a better understanding of pollutants and to develop treatment and disposal processes for dealing with them. Increased attention should be given to the following areas (after National Research Council, 1987, pp. 164–168):

- Manufacturing processes and design: research directed toward the cost-effective alleviation of environmental hazards arising from the manufacturing industries
- Combustion: increased fundamental understanding of the physics and chemistry of combustion in order to develop improved incinerator technology (involving thermal processes, incineration, pyrolysis, biological and wet combustion processes) and methods for control of hazardous emissions
- Microbial transformation: basic knowledge about microorganisms, their physiology, biochemistry, and ecology, to develop further the biotechnology for transforming dilute hazardous waste
- Assimilative capacity of the global environment: research directed toward understanding the movement, fate, and effects of chemicals in the environment in order to develop control strategies that make more effective use of the ability of the environment itself to deal with contaminants
- Sensors and measurement methods: development of improved sensors to gather more comprehensive information and of analytical modeling techniques that can integrate this information and identify viable control strategies

Several specific areas likely to have significant effects on our ability to deal with long-term environmental changes relate to energy: electric powered vehicles; reduction of sulfur dioxide, oxides of nitrogen, and particulate emissions; improvements in electric energy use; energy storage; reduction of power requirements; and fuel cells. Topics that have long been on the environmental research agenda, such as recycling or reuse and the development of biodegradable materials, remain largely unsolved problems still requiring attention.

## LEAVING THE END-OF-PIPE APPROACH

On the regulatory side, the Environmental Protection Agency (EPA) has been properly criticized for strategies or policies used to address pollution problems that focus almost exclusively on "end-of-pipe" solutions to pollution problems. Such practices focus almost exclusively on treating what comes out of the pipe or smokestack, ignore broader systems-oriented approaches and the assimilative capacity of the environment, impose lockstep application of the "best available technologies," and thus hinder innovation. Embedded in such policies are disincentives that hinder our capacity to address these problems at more efficient and productive levels, such as waste reduction and prevention, recycling and reuse, isolation of wastes, and substitution of materials in manufactured goods. To address the global and local changes in the environment, fundamental changes must be made in the strategies we have been pursuing, as an EPA Science Advisory Board Report recently urged (U.S. EPA, 1988). A careful reexamination of the targets for technology is required, especially if engineering and the applied sciences are to be effective in addressing environmental changes that are already well understood and accepted, as well as those that appear to be emerging.

Although one ought to be impressed with the political acumen of those who have used the events of the summer of 1988 to increase the awareness of changes in the global environment, few believe that doomsday is in sight (Solow, 1988). However, most of us will never know whether they are right or wrong. The consequences to ourselves and, more important, to future generations are so monumental that it would be irresponsible not to face up to these matters. The crisis before us is of a special kind: it demands the rejection of avoidance and denial, and a genuine and complete national commitment to confront the environmental changes that lie before us.

## ACKNOWLEDGMENTS

I want to thank James Coulter, the late Abel Wolman, and Daniel Okun for taking the time to read an earlier draft of this chapter and provide me with their thoughtful and helpful comments. I owe special thanks to Mike Lynn and Judy Bowers for their editorial help. Needless to say, all of them are absolved from responsibility for the final product.

## NOTES

1. These kinds of episodes are likely to continue even though Resource Conservation and Recovery Act rules do not permit any waste defined as hazardous to be exported "unless . . . EPA . . . and the government of the country [involved] . . . consent in writing to accept the waste" (Bernthal, 1988).
2. Although poorer nations do not relish this kind of trade, they may find the offer of cash for accepting hazardous wastes an acceptable trade-off in the short run (Shabecoff, 1988).
3. Incineration is an "old" technology for which there has been some innovation recently, including refuse-derived fuels, improved burning techniques and boiler design, fly ash handling and disposal, etc.
4. It is important to recognize the link between our environmental problems and the quality of our infrastructure. Much of the U.S. urban infrastructure is already recognized as inadequate, and most municipalities face a crisis brought about by decayed, ineffective, and inoperative urban services. *Newsweek* estimated that it would cost approximately $3 trillion dollars to put these systems back in working order. The difficulties of raising such vast sums of money for sewers, water pipes, roads, bridges, and the like, truly represent a crisis of major proportions (National Council on Public Works Improvement, 1988).
5. A recent report concludes that atmospheric circulation anomalies were the primary cause of the drought of 1988, not greenhouse warming. "Any greenhouse gas effects may have slightly exacerbated these overall conditions . . . but they almost certainly were not a fundamental cause" (Trenberth et al., 1988).
6. The 1987 budget allocation for resources and the environment decreased by $123 million, or 12 percent, after remaining almost constant in 1986. This category involved expenditures representing 1.5 percent of the total federal R&D budget; expenditures for waste treatment and disposal R&D represent a tiny fraction of that total.

## REFERENCES

Bernthal, F. M. 1988. U.S. views on waste exports. U.S. Department of State, Current Policy No. 1095. Washington, D.C.
Evans, H. M. July 11, 1988. Editorial. U.S. News and World Report. P. 67.
Hardin, G. 1968. The Tragedy of the Commons. Science 162:1243–1248.
Kahn, A. E. 1966. The tyranny of small decisions: Market failures, imperfections and the limits of economics. Kyklos 19:23–24.
National Council on Public Works Improvement. 1988. Fragile Foundations: A Report on America's Public Works. Washington, D.C.: U.S. Government Printing Office.
National Research Council. 1984. Disposal of Industrial and Domestic Wastes: Land and Sea Alternatives. Board on Ocean Science and Policy, Commission on Physical Sciences, Mathematics, and Resources. Washington, D.C.: National Academy Press.
National Research Council. 1987. Directions in Engineering Research: An Assessment of Opportunities and Needs. Engineering Research Board, Commission on Engineering and Technical Systems. Washington, D.C.: National Academy Press.
Public Works. 1988. Editorial viewpoint. Public Works 118(8):7.
Shabecoff, P. July 5, 1988. Irate and afraid, poor nations fight efforts to use them as toxic dumps. New York Times 137:22(N), C4(L).

Solow, A. R. December 28, 1988. Pseudo-scientific hot air: The data on climate are inconclusive. New York Times 138:A15(N), A27(L).
Trenberth, K. E., G. W. Branstator, and P. A. Arkin. 1988. Origins of the 1988 North American drought. Science 242:1640–1645.
U.S. Environmental Protection Agency. 1987. Unfinished Business: A Comparative Assessment of Environmental Problems. Office of Policy Analysis. (NTIS-PB88-127048). Washington, D.C.: U.S. Government Printing Office
U.S. Environmental Protection Agency. 1988. Future Risk: Research Strategies for the 1990s. Science Advisory Board. (NTIS SAB-EC-99-040). Washington, D.C.: U.S. Government Printing Office.
Weinberg, A. M. 1966. Can technology replace social engineering. Bulletin of the Atomic Scientists 22(10):4–8.

# The Paradox of Technological Development

## PAUL E. GRAY

Technological development has had profound and permanent effects the way we live and the way we think about the future—what is possible, what is probable, what is to be feared, and what is to be hoped for. It also provides an appropriate introduction to my principal theme, which is a paradox of our time: the mixed blessing of almost every technological development. Technological developments come about as people seek solutions to specific problems and needs, and they often open the way to other innovations and applications that were unimaginable at the outset. Because we have not been able to predict all of their consequences, nearly all such developments carry with them the potential for misuse, and many consequences are rightly regarded as not only unfortunate but also malign in their impact.

The new ideas and technologies resulting from the efforts of engineers are, in some respects, like the Golem of the Rabbi of Prague. An artificial creature, created to serve, the Golem exhibited a mind of its own, acting in mischievous ways unanticipated by its maker. New technology will be applied in ways that transcend the intentions and the purposes of its creators, and new technology will reveal consequences that were not anticipated.

Consider, for example, the "green revolution." Developments in agriculture have improved food production around the world. Countries such as India, which for decades was unable to feed its people, have become net exporters of food. At the same time, growing reliance on insecticides and fertilizers has contributed to widespread chemical pollution of rivers, lakes, and seas, threatening the food chain itself. Other examples abound: the automobile, mass communications, energy production. All have changed our lives for the better, and all have consequences that threaten our well-being as individuals and as a global society.

What is it about technological development today that makes it such a mixed blessing and leads to such widespread wariness on the part of the public? How can this double-edged quality of technological development be understood in ways that will help us avoid some of the pitfalls of the past? What can we do about engineering education, engineering practice, and public policy to help resolve the paradox and reduce the chances of creating new problems in the future?

## SOME CHARACTERISTICS OF TECHNOLOGICAL DEVELOPMENT

Let me begin with some of the characteristics of technological development that have caused us problems in the past—both in practice and in perception.

First, major new technological developments produce changes that deeply affect society and do so in ways that make it impossible to contemplate turning the clock back by rejecting the development. The very power and perceived permanence of new technology surely contribute to the wariness with which it is regarded by many; the green revolution is a good example. Although new technologies can be adapted to address some of the unfortunate consequences of modern agricultural methods, a wholesale abandonment of those methods is now unthinkable; it would lead to malnutrition and starvation on a scale unknown in human history.

Second, more recent technological developments are, in many cases, incremental in their intended beneficial consequences. This may have been less frequently the case in the earlier stages of development when the benefits of a new technology, such as electrical energy distribution systems, were dramatic in their effect. Increasingly, the positive consequences of a development are, or are understood as, incremental or marginal in character. As a result, the natural human tendency to avoid change, the unknown, and risk becomes more dominant in considering new technological developments.

A third characteristic of technological development relates to our steadily improving ability to quantify very small amounts of potentially hazardous materials in our environment, as well as our continually changing assessment of hazards and degree of risk. For example, when DDT was developed a half-century ago, it led to dramatic reductions in the incidence of malaria and was hailed as a great benefit to humankind. Since then, our growing ability to identify and measure very small concentrations of this and other synthetic pesticides has enabled us to recognize the harm they do to our environment as well.

Even with our growing capabilities to identify and measure hazards, when it comes to questions of probability, uncertainty, and long-term consequences, scientists disagree among themselves about the bases of risk assessment. Policymakers do the best they can, but when they get many different opinions from experts, it becomes just that much more difficult to know what to do. This certainly does not help public understanding and debate on such issues.

Fourth, because we now live on a crowded planet, the consequences of technological development have a more immediate and far-reaching impact and are more readily apparent than in earlier times. For most of human history, the impacts of development were masked and diluted because that development was orders of magnitude away from stretching the capacity of our environment to absorb pollution and other burdens. Our heritage in this respect has roots that go a long way back. The slash-and-burn agriculture of prehistoric humankind required new land every few years, but this was surely never seen as an obstacle or consideration because land was available without apparent limits and the people were so few. Air pollution in industrial England had severe local effects, as in the killing smogs of London, but these problems seemed not to have significant global consequences and were, in any case, largely dealt with locally. The impact of technological development on our environment as reflected in degraded air and water quality, warnings of possible global warming and the depletion of stratospheric ozone, and the hazards of toxic waste is, in large measure, a consequence of the fact that there are many more of us on this planet. Consequences that were unimportant—even practically undetectable—when the earth sustained 1 billion or 2 billion humans become dangerous, or even intolerable, when there are 5 billion, rapidly heading toward 10 billion.

## PUBLIC PERCEPTIONS AND PUBLIC POLICY

In addition to these characteristics of technology itself, the paradox of technological development is compounded by public perceptions about risk and by the fact that we lack an effective system for developing public policies to help guide technological development, particularly as we face these issues as a global society.

The public perception of risk is sometimes unpredictable and inconsistent with quantitative risk assessment data. For example, the public tolerates approximately 50,000 deaths a year on our nation's highways with no great outcry, yet there is widespread public concern each time a plane crashes. Although statistics show that air travel is much safer than auto travel, the public perception is different. Both scientific literacy and communication about risk should be improved so that individuals are better educated and public perception is closer to the quantitative realities. At the same time, we must understand and be sensitive to public perceptions even if they do not appear to be consistent with quantitative evidence. We should also recognize that even a public educated in the most precise and sophisticated risk assessment techniques will distrust policymakers and scientists or engineers who dismiss its legitimate fears.

As for the development of public policy, government and industry in this country have had a largely adversarial relationship when it comes to policies regarding environmental and economic consequences of technological development, with a reliance on regulation rather than on cooperation. This stems largely from the lack of clearly articulated and agreed upon standards for safety, cleanliness, or risk. Without such criteria, it is not surprising that continual conflict and misunderstanding persist between groups and individuals with differing concerns. This, together with the lack of technical and scientific knowledge at high levels of decision making in the legislative and executive branches of government and in the public itself, has meant that we do not have a consistent, well-thought-out, and clearly articulated set of policies in this domain. Nor do we have processes that allow us to resolve disputes in a reasonable fashion. We are caught in a gridlock of adversarial relations among various special interest groups, a position that exacerbates the problem rather than helps resolve it.

Creators of new technological developments and policymakers thus have a particular responsibility to explore, as thoroughly and aggressively as possible, the multiple consequences of new developments to make those considerations an integral part of the process of technological development. They need to develop guidelines and policies for sustainable development that reflect concern for the long-term, global implications of large-scale technologies and that support the innovation of less intrusive, more adaptable technologies at all levels.

## A CASE IN POINT: THE GREENHOUSE EFFECT

A dramatic and current example of the double-edged quality of technological development and of a problem that will require the most concerted technological, political, economic, and social collaboration on an international basis is the phenomenon known as the greenhouse effect—an environmental issue that suddenly moved toward the head of the list of public concerns following the unusual weather conditions in the United States during the summer of 1988.

The average temperature of the earth manifests a balance between the heating effects of solar radiation and the cooling associated with infrared thermal radiation from the earth. The atmosphere's transparency and, therefore, the average global temperature depend on the atmosphere's absorption characteristics and concentrations of carbon dioxide and certain trace gases. Some of these gases are produced by natural processes and have been present in our atmosphere for eons. Because of their presence, the earth is about 30°C warmer than it otherwise would be; this is the phenomenon known as the greenhouse effect.

These gases are also produced by industrial processes, particularly (for carbon dioxide and nitrous oxide) by the burning of wood and fossil fuels, and there is now clear evidence that the concentrations of these greenhouse gases are steadily increasing. As a result, the average temperature of the earth must increase to maintain the heat balance between solar input and infrared output.

The growing concentrations of greenhouse gases in our atmosphere due to industrial, agricultural, and other human activities are, in a sense, directly driven by population and by the increasing intensity of development in its present form. The earth has experienced an increase in atmospheric carbon dioxide of about 20 percent in less than two centuries. The present rate of increase suggests that the concentration of this greenhouse gas will increase another 10 percent by early in the next century and will double by the second half of the twenty-first century (Bolin et al., 1986; Trabalka, 1985).

Changes in the atmosphere during the past century should, according to theoretical models, have produced about 0.5°C of warming. Whereas the average global temperature has increased by about this amount, the natural variations in temperature are of about the same order of magnitude. Consequently, direct confirmation of the global warming associated with increases in greenhouse gases is not yet in hand, although the observed temperature increases are consistent with theoretical expectations. These theoretical models predict an additional 0.5°C warming in the next 20 years and 2–5°C warming by the middle of the twenty-first century if greenhouse gas concentrations continue to increase as energy use increases and as deforestation continues.

The warming is buffered and delayed by the oceans, which absorb both carbon dioxide and heat. As a consequence of this, even if production of greenhouse gases could be stopped dead today, global temperatures would continue to increase for several decades. These temperature increases will persist because most of the greenhouse gases have very long lifetimes. Carbon dioxide is removed from the atmosphere by two processes: net photosynthesis in plants and absorption in the oceans, with eventual deposition at the ocean bottom as limestone. The second process is both dominant and extremely slow, with time constants on the order of 1,000 years.

Although direct evidence of global warming attributable to greenhouse gases has not yet been obtained, our present understanding of the mechanisms and our direct observation of increasing greenhouse gas concentrations make eventual significant warming a virtual certainty. It is likely that the earth will, by the end of this century, be warmer than it has been in the past 100,000 years. Unless we change course, global temperatures are likely to be higher by the latter half of the twenty-first century than they have been in 2–10 million years.

The effects of global warming on climate and thus on the activities of humankind are much harder to predict. Increases in sea level are inescapable as the warming oceans expand and as mountain glaciers and ice caps release water. Patterns of precipitation are likely to change, thus bringing less rainfall in the middle latitudes where much of the world's grain production now occurs; the viability and reproductive capacities of plants of all kinds, particularly unmanaged forests, could diminish. Those extreme natural events, which cause much human misery—drought, heat waves, coastal flooding—are likely to become much more frequent.

## ADDRESSING GLOBAL WARMING AS A GLOBAL PROBLEM

What should be done about this? The problem of global warming calls for both human adaptation and the limitation of pollutants. Each requires technological support and engineering development, and both require cooperation—not only among those in the engineering profession, industry, and government but also among nations.

Adaptation in anticipation of a warmer earth is necessary because the most drastic course of limitation of pollutants will not offset the momentum of past contamination; a significant degree of warming is now unavoidable. Adaptation will require attention to agriculture, including the development of new strains of grain, to water resources, and to protection of low-lying coastal regions where flooding will occur.

Limitation is essential if we, as a global population, are to avoid even more extreme conditions far into the future. Further, limitation of greenhouse gases can slow the rate of warming, which eases somewhat the task of adaptation.

The United Nations Environment Program (UNEP) and the World Meteorological Organization recently recommended the following actions to reduce carbon dioxide emissions in the face of growing populations and increased economic activity (Jäger, 1988):

- Reduce fossil fuel use by increasing end-use energy efficiency. The experience of the past 15 years in response to the increase in oil prices induced by the Organization of Petroleum Exporting

Countries provides us with an example of the power of conservation. The UN study foresees a potential reduction in energy consumption in the industrialized nations of 50 percent with extant technology. Many efficiency improvements can be achieved with net economic savings, and conservation efforts can be undertaken right now, without delay.

- Shift the fossil fuel mix from high carbon dioxide-emitting fuels to those that produce less carbon dioxide per unit of energy. Natural gas is better than oil, which is better than coal.
- Reverse current trends toward deforestation and encourage reforestation.
- Develop the technology to remove carbon dioxide from stack gases of large, stationary fossil fuel-burning energy converters, such as electric power generating plants, and dispose of it in the deep ocean. Although such an approach would at least double the cost of electricity, these costs are about the same magnitude as those associated with the most stringent pollution control requirements now in place in some nations.
- Replace fossil fuels with alternative energy sources such as solar energy, wind and tidal power, ocean thermal conversion, and nuclear power. This is, to my mind, the only viable long-term approach to offset the forces of continued population and economic growth.

It is clear that we are not talking simply about technological solutions. Global warming is, obviously, a global problem; any effort to limit future emissions of greenhouse gases must be global in character if it is to be effective. The degree of cooperation required is without precedent because it must encompass both the highly industrialized nations, where present energy use is most intensive, and the less developed nations, where hopes for a better future appear to require greater intensity of energy use. For example, what response should the West expect from China if we, who have contributed most of the present carbon dioxide buildup in the atmosphere, suggest that the Chinese, in the interest of a less degraded environment a century from now, should forgo the exploitation of their enormous reserves of coal?

Our traditional political processes tend to deal with near-term issues and immediate problems. We must develop political processes capable of producing sensible responses to problems where the time constants are on the order of a century.

## ONE COURSE OF ACTION: NUCLEAR ENERGY

Whereas economic, political, and social forces must be brought to bear on this problem, it seems self-evident that amelioration of this problem requires new engineering creativity and technical developments aimed at the several courses of action described in the UN study. Although this is not the place for a careful exploration of possible future developments and directions, I would like to comment briefly on one aspect of amelioration, which seems to me to be compelling: the greater use of nuclear energy as an alternative to fossil fuels.

It has become a commonplace to assert that the nuclear industry in the United States is now dead, that its death was probably suicide, and that the public is both passionate and unified in its determination to see that it stays buried. The present state of affairs needs no explication. Three Mile Island and Chernobyl cannot be expunged from our collective consciousness. Seabrook and Shoreham are real-time examples of the depth of the conviction held by our political leaders, perhaps even by a majority of the public, about the risks and benefits of nuclear power.

Certainly, mistakes have been made in the past, both in technology and in the ways public concerns about nuclear energy have been addressed. We must be willing to learn from these mistakes, to explore different approaches to the design of nuclear energy plants, and to improve public awareness and understanding of these issues if nuclear energy is to play a role in our future.

Let me speak first about reactor design. Light-water reactors (LWRs), which are used in nearly all of the plants in operation or under construction in the United States, place heavy demands on the builders and operators of these plants. The principal safety hazard is a loss-of-cooling accident, which could lead to the melting of fuel elements and subsequent release of radioactivity. To prevent such an occurrence, the design and operation of an LWR must provide an absolute guarantee of the presence of adequate quantities of cooling water, and the guarantee must reflect the worst possible scenarios, including rupture of pipes, pump failures, failure of outside electrical power, and operator errors. To reduce the probability of loss of coolant to acceptably small levels, LWRs rely on multiple redundant backup systems or "defense in depth." It is the nature of these complex and tremendously costly protective systems that their effectiveness under all accident conditions cannot be demonstrated experimentally. Consequently, questions about modes of failure can be answered, at best, only in analytical and probabilistic terms, which is a major reason for much of the public skepticism about nuclear power in its present form.

It is possible to design and build reactors that can survive the failure of components without the possibility of fuel damage or the release of radioactivity. This can be accomplished by employing forms of fuel able to withstand very high temperatures, by limiting the power density in the core, and by arranging for sufficient heat removal by natural processes to prevent fuel damage. Such "passively safe" reactors can be designed to suffer simultaneous failure of all control and cooling systems without endangering the public (Agnew, 1981; Faltermayer, 1988; Lidsky, 1988).

Reactors designed in this manner produce less power output than light-water reactors: 100–150 megawatts of electrical power output compared with 1,000–1,500 megawatts. A number of individual power-producing modules will be combined on each site to produce the required amount of power. These small, identical modules can be factory built instead of being custom made on-site, as is the case for the much larger, much more complex LWRs. The economy of serial production will replace the economy of large scale.

Because the individual reactor modules are identical and centrally built, licensing can be standardized and can be based on full-scale testing of an actual device rather than on detailed review and inspection of the defense in depth required for LWRs. This is an enormous advantage because it permits actual demonstration of the response of the reactor to the most severe and demanding hazards. Reactors of this kind present a vanishingly small operating risk to the public—a risk much smaller than that associated with most everyday activities. Coal-fired electric power plants produce and release more low-level radioactivity (carried in fly ash) than do nuclear reactors (Hurley, 1982).

Public attitudes about the acceptability of nuclear power are based as well on concerns about high-level nuclear waste handling and disposal. Decades of temporizing and indecision in the United States have aggravated this problem. What is required here is not simply technical innovation, but political creativity as well, to address the dilemmas posed by the "not in my backyard" concerns. Several nations in Western Europe have shown that solutions to this problem do indeed exist.

I am convinced that several undertakings are essential if nuclear power is to have any role in the U.S. energy future:

1. We must make an earnest and sustained effort to educate the public about the risks and benefits of nuclear power in terms that permit quantitative comparison with other energy sources.

2. We must achieve technically, politically, and environmentally acceptable solutions to the problem of nuclear waste handling and disposal—solutions that take into account the associated concerns about nuclear weapons proliferation.

3. We must develop, build, and test radically different reactor designs that pose negligible risks of the accidental release of radioactive materials

as a result of overheating. Several possibilities exist, including new water-cooled and liquid-metal-cooled designs as well as gas-cooled designs. These designs hold the promise of passively safe operation.

Nevertheless, it is clear that none of these designs will be acceptable until such reactors are built at scale and thoroughly tested under the most extreme conditions Murphy's Law can produce. Absolutely risk-free operation cannot, like absolutely anything else, be guaranteed: one can postulate a meteor falling on the reactor, after all. Nevertheless, the degree of risk to the environment and to human life can be driven down below the levels of corresponding risk inherent in the present alternative of fossil fuels.

## ENGINEERING EDUCATION AND PRACTICE: WHAT NEXT?

Richard de Neufville, chairman of the Technology and Policy Program at the Massachusetts Institute of Technology, has suggested that many people who seek solutions to complex, important issues, such as toxic waste, nuclear power, or global warming, tend to reflect one of two perspectives: some assert that every problem has a technical "fix"; others, that each of these same problems has a moral fix.

Unfortunately, neither perspective admits to the complexities and to the social, technical, and moral implications of most important, real problems. Silver bullets exist only in myths, and responsible solutions are developed only by knitting together the technical and moral perspectives. Those of us who develop, promote, and apply technological innovation have the moral responsibility to explore and consider, to the greatest extent possible in the light of our best effort, the full consequences of any innovation. It is both professionally and morally irresponsible to define the problem so narrowly as to leave these considerations to others.

What can be done in engineering education and practice, and in the domain of public policy, to recognize this conflict between the potential and the problems of technological development, to deal realistically with public apprehension about the risks attendant on change, and to minimize the degree to which future developments are burdened with unforeseen negative consequences?

With regard to engineering education, a number of things could be done:

1. Instruction in the humanities, arts, and social sciences should be structured and undertaken to require the engineering student to gain some understanding of societies and cultures, of the complex relationships between society and technology, and of human values and relationships.

Engineering is, obviously, a socially derived and culturally influenced activity, and engineers cannot function effectively without being steeped in those contexts. This is not the only reason for studying the humanities, arts, and social sciences—to make better engineers. However, an engineer who has cultivated an interest in one or more of these fields is, I believe, more likely to bring to his or her practice a sensitivity to the social context of engineering and attention to all the consequences of new technology.

2. Although all engineers should have an appreciation for and sensitivity to the social environment in which they operate, some engineers—who might be called interface engineers—will work directly on issues of application, impact, and implementation in a broader context. They need direct engagement with these issues in their education. These students should tackle subjects and engage in research on topics that directly address the political, economic, and social considerations integral to scientific and technological developments.

3. Engineering design courses, particularly at the upper level, should move beyond requiring significant individual effort to requiring collaboration among teams of students formed to work on problems that are not artificially isolated from their social context. A part of that team effort should bear on the exploration of social consequences and the problems that arise when technology is used for different purposes. Although such projects are inevitably constrained, it is important that engineering students begin to work as engineers in ways that reflect to some degree the way actual engineering work should be done.

4. Finally, students should be prepared for active leadership in the definition and resolution of issues that arise at the intersection of technology and society. Neither we nor they can afford to sit back and expect other professions to imagine, create, and implement the kinds of solutions that are both socially responsible and firmly grounded in technical realities. Engineers do not hold the sole responsibility here, but the profession must consciously prepare and train itself to do its part: effective leadership must be learned.

Now this speaks as well to the role of the engineer. Engineering practice must, in the work of the engineer, reflect a broadened role and more comprehensive concerns. The engineer should bring to his or her work not only sound technical knowledge, disciplined technique, and a focused search for creative solutions to novel problems but also a concern for the ecology of technology. Not all consequences of technological development can be anticipated; not all unfortunate extensions can be anticipated. Nevertheless, the imperative to understand the implications of a development in its broadest and most encompassing terms is a professional responsibility of the engineer, which must be incorporated into the task from the outset.

On the other hand, I am not suggesting that this responsibility rests with the engineer alone. The engineering profession should not only incorporate social and economic considerations into its work but also work together with government, industry, and the public to develop long-term, global strategies for addressing these issues.

## COOPERATION FOR THE FUTURE

The issues raised by the paradox of technological development are profound and difficult. Nonetheless, I am optimistic that, in this era of global interdependence, responsible people will recognize that appropriate public policies to ensure sustainable development can and must be developed from an international perspective. We should begin now to lay a firm foundation for the future.

In particular, the following challenges should be considered:

- to educate engineers to consider the far-reaching implications of their work for the social and physical environment, and also to educate those in the humanistic disciplines to fully appreciate the nature of science and engineering;
- to develop technology for sustainable development, appropriate allocation of resources, and risk management;
- to advance the art of policymaking at all levels, which includes realistically reflecting the implications of technological innovations in both the substance of decisions and the process of decision making; and
- to recognize the need for communication, firm resolve, and mutual respect among policymakers, engineers, industrialists, and the public, who will ultimately be responsible for our common future in the democracies.

Most important, we should not let the need for adequate preparation be an excuse for inaction. Just as long lines at the gas pumps in the winter of 1973–1974 triggered public awareness of the need for conservation and alternative energy sources, the hot, dry summer of 1988 may inspire the search for technologies and public policies that respect the limitations of the environment and allow for economic growth.

## CONCLUSION

Engineers have changed the world we live in. Engineers with vision can provide the means to realize strategies for a viable future in our economically, culturally, and ecologically intertwined world.

The great hope and the great challenge before us are to bring engineering education and practice, industrial priorities, and public policy into alignment in ways that eliminate the paradox of technological development. We have an opportunity now to turn that paradox around and forge a new concept of how the engineer works and views the world. Furthering technological and economic development in a socially and environmentally responsible manner is not only feasible, it is the great challenge we face as engineers, as engineering institutions, and as a society.

## ACKNOWLEDGMENTS

I am grateful to the following persons for discussions that were helpful in the preparation of these remarks: Hermann Haus, Lawrence Lidsky, Kathryn Lombardi, Richard de Neufville, Ronald Prinn, Daniel Roos, Walter Rosenblith, Peter Stone, Neil Todreas, Leon Trilling, Robert White, and Gerald Wilson.

## REFERENCES

Agnew, H. M. 1981. Gas cooled nuclear power reactors. Scientific American 244:55–63.
Bolin, B., B. R. Döös, J. Jäger, and R. A. Warrick. 1986. The Greenhouse Effect, Climatic Change, and Ecosystems. New York: John Wiley & Sons.
Faltermayer, E. 1988. Taking fear out of nuclear power. Fortune 118(1 August):105–118.
Hurley, P. M. 1982. Living with Nuclear Radiation. Ann Arbor, Mich.: University of Michigan Press.
Jäger, J. 1988. Developing Policies for Responding to Climatic Change. WCIP-1, WMO/TD-No. 225, April. Geneva: World Meteorological Organization and United Nations Environment Program.
Lidsky, L. M. January 10, 1988. A safe atomic plant for the future? Washington Post C3.
Trabalka, J. R. 1985. Atmospheric Carbon Dioxide and the Global Carbon Cycle. DOE/ER-0239. Washington, D.C.: U.S. Department of Energy.

# Contributors

SIAMAK A. ARDEKANI is assistant professor of civil engineering at the University of Texas in Arlington. His current research is in transportation management issues in the aftermath of major urban disasters such as earthquakes and floods. He has coauthored numerous journal articles on urban traffic management and operation, and is associate editor of the *Transportation Science Journal*. He serves on the Committee on Traffic Flow Theory and Characteristics organized by the Transportation Research Board of the National Research Council. Dr. Ardekani received his Ph.D. degree in civil engineering from the University of Texas at Austin.

JESSE H. AUSUBEL is a fellow in science and public policy at The Rockefeller University in New York City and director of studies for the Carnegie Commission on Science, Technology, and Government. From 1983 through 1988 Mr. Ausubel served as director of the Program Office of the National Academy of Engineering. Mr. Ausubel first came to the Academy complex as a resident fellow of the National Academy of Sciences in 1977. He then served for two years as a research scholar in the resources and environment area at the International Institute for Applied Systems Analysis in Laxenburg, Austria. From 1981 to 1983 he served as a National Research Council staff officer principally responsible for studies of the greenhouse effect. Mr. Ausubel is author or editor of numerous publications in the field of climatic change. Among his current areas of research are calculation of industrial emissions to the atmosphere, long-term interactions of environment and technology, and comparative diffusion of technologies in different countries.

ROBERT U. AYRES is professor of Engineering and Public Policy at Carnegie Mellon University in Pittsburgh, Pennsylvania, and deputy leader of the Technology-Economy-Society Program at the International Institute of Applied Systems Analysis, Laxenburg, Austria. He is author or coauthor of nine books and many journal articles, as well as numerous book chapters, symposium papers, and technical reports on a variety of subjects. He is a "futurist" as well as a systems analyst. His current research is focused on technological change, with special emphasis on computer-integrated manufacturing and "industrial metabolism." Dr. Ayres received his B.S. degree in mathematics from the University of Chicago, an M.S. in physics from the University of Maryland, and a Ph.D. in mathematical physics from Kings College, University of London.

RICHARD E. BALZHISER is president and chief executive officer of the Electric Power Research Institute (EPRI) located in Palo Alto, California. Dr. Balzhiser joined EPRI in 1973 as the director of the Fossil Fuel and Advanced Systems Division and served in several senior executive positions before becoming EPRI's executive vice president in 1987. From 1971 to 1973 he served as an assistant director of the White House Office of Science and Technology Policy, where he led energy, environment, and natural resource activities. Previously, Dr. Balzhiser was chairman of the Department of Chemical Engineering at the University of Michigan in Ann Arbor, except for 1967–1968 when he served as a White House Fellow in the office of the Secretary of Defense. Dr. Balzhiser currently serves on the advisory boards of the Institute for Energy Analysis, the University of Michigan College of Engineering National Advisory Committee, and the Academy Industry Program of the National Academy of Sciences, National Academy of Engineering, and Institute of Medicine. He was recently appointed to the U.S. Department of Energy's Innovative Control Technology Advisory Panel. Dr. Balzhiser received his B.S. and Ph.D. degrees in chemical engineering, and his M.S. degree in nuclear engineering from the University of Michigan.

SHELDON K. FRIEDLANDER is Parsons Professor of Chemical Engineering and director of the newly established Engineering Research Center for Hazardous Substance Control at the University of California, Los Angeles. Dr. Friedlander's research has involved air quality engineering and aerosol technology, the behavior and characterization of particulate matter in gases and liquids, and air quality/emission source relationships for particulate pollution. From 1984 to 1988 he chaired the UCLA Chemical Engineering department, and before that he was professor of chemical engineering and environmental engineering at the California Institute of Technology. He has consulted for the Los Angeles Air Pollution Control District, and served as chairman of the National Research Council Panel on the Abatement of Particulate Emissions from Stationary Sources as well as the subcommittee on Photochemical Oxidants and Ozone. He was also chairman of the Environmental Protection Agency Clean Air Scientific Advisory Committee and is a member of its Science Advisory Board Executive Committee and of the National Academy of Engineering. Dr. Friedlander received his B.S. degree in chemical engineering from Columbia University and Ph.D. degree from the University of Illinois.

ROBERT A. FROSCH is vice-president in charge of General Motors research laboratories. Dr. Frosch's career combines varied research and administrative experience in industry and in government service. He has been involved in global environmental research and policy issues at both the national and the international level. From 1951 to 1963 he was employed at Hudson Laboratories of Columbia Univeristy, first as a research scientist and then as director from 1956 to 1963. In 1963 he became director for Nuclear Test Detection in the Advanced Research Projects Agency (ARPA) of the Department of Defense, and deputy director of ARPA in 1965. In 1966 he was appointed assistant secretary of the Navy for research and development. He served in this position until January 1973, when he became assistant executive director of the United Nations Environment Program. In 1975 he became associate director for applied oceanography at the Woods Hole Oceanographic Institution, and from 1977 to 1981 he served as administrator of the National Aeronautics and Space Administration. He served as president of the American Association of Engineering Societies from 1981 to 1982. Dr. Frosch is a member of the National Academy of Engineering. He received his A.B., M.S., and Ph.D. degrees in theoretical physics from Columbia University in New York.

JOSEPH P. GLAS is director of the Freon Products Division for the Chemicals and Pigments Department at the Du Pont Company. He joined Du Pont in 1964 as a research engineer at the company's Circleville, Ohio, research and development laboratory. He became product manager of Kapton polyimide and metallized Mylar polyester in 1974, manager of packaging market development at the Chestnut Run Technical Service Laboratory in 1975, and research manager for the commercial resins division in 1976. Dr. Glas became research director of the Atomic Energy Division of the Petrochemicals Department at its Savannah River Laboratory in 1979, and from 1980 to 1982 he served as research director for Remington Arms Company, a wholly owned subsidiary of Du Pont. He became director of research and development for the Chemicals and Pigments Department in 1982, and assumed his present position in 1985. Dr. Glas received a

B.A. in chemistry from Rockhurst College in Missouri and M.S. and Ph.D. degrees in chemical engineering from the University of Illinois.

PAUL E. GRAY is president of Massachusetts Institute of Technology. Before becoming president, he served on the faculty and in the academic administration, notably as associate provost, dean of engineering, and chancellor. The author or coauthor of numerous basic texts in electrical engineering, Dr. Gray's professional interests are in semiconductor electronics and circuit theory. As a member of the faculty, Dr. Gray won recognition for his teaching and for his contributions to the revitalization of engineering education. In recent years, he has been a leader in the continuing development and reshaping of the undergraduate curriculum. Under Dr. Gray's administration, the university has expanded its relations with industry, both in this country and abroad, and is developing major research and education programs in such areas as microelectronics, health sciences and technology, communications, the brain and cognitive sciences, and the management of technology. Dr. Gray is a member of the National Academy of Engineering. He earned his S.B., S.M., and Sc.D. degrees in electrical engineering from the Massachusetts Institute of Technology.

ROBERT HERMAN is L.P. Gilvin Centennial Professor, Emeritus, in civil engineering and sometime professor of physics at the University of Texas at Austin. Before assuming his present position in 1979, Dr. Herman was with the General Motors Research Laboratories and headed the Department of Theoretical Physics from 1959 to 1972 and the Traffic Science Department from 1972 to 1979. Dr. Herman's research has covered a wide range of both theoretical and experimental investigations, including molecular and solid-state physics, high-energy electron scattering, astrophysics and cosmology, as well as operations research, especially vehicular traffic science and transportation. With Ralph Alpher in 1948, Dr. Herman made the first theoretical prediction that the universe should now be filled with a cosmic microwave background radiation, which is key evidence for the validity of the Big Bang model of the origin of the universe. Dr. Herman is a member of the National Academy of Engineering. He received his B.S. degree in physics at City College, New York, and his master's and Ph.D. degrees in physics from Princeton University.

THOMAS H. LEE is professor of electrical engineering at the Massachusetts Institute of Technology. His interests include electric power systems engineering and physical electronics, energy technology and policy, and technology assessment and strategic planning. In 1948 he began work with General Electric where, over the course of 32 years, he held numerous posts from senior research engineer (1955–1959) to staff executive and chief technologist (1978–1980). In 1980 he left General Electric to become director of the Electric Power Systems Engineering Laboratory and Philip Sporn Professor of Energy Processing at the Massachusetts Institute of Technology. In 1984 he became director of the International Institute for Applied Systems Analysis in Laxenburg, Austria, for a three-year term. He is a member of the National Academy of Engineering. Dr. Lee received his doctorate in electrical engineering from Rensselaer Polytechnic Institute.

WALTER R. LYNN is professor of civil and environmental engineering and dean of the university faculty at Cornell University. His teaching and research have focused on applying analytical methods to public policy decisions composed of technical, political, social, and economic elements. In 1961 he joined the Civil Engineering Department at Cornell to develop a graduate program in environmental systems engineering. He served for eight years as director of the Cornell Program on Science, Technology, and Society. Before that he was Director of Cornell's Center for Environmental Research and director of the School of Civil and Environmental Engineering. Dr. Lynn has served as associate editor of the *Journal of Operations and Research* and the *Journal of Environmental Economics and Management*. He was the first chairman of the Water Science and Technology Board of the National Research Council, and is currently a member of its Committee on Water Resources Research. He also serves as Governor Cuomo's appointee as chairman of the New York State Water Resources Planning Council. Dr. Lynn received his B.S. degree in civil engineering from the University of Miami, an M.S. in sanitary engineering from the School of Public Health at the University of North Carolina, and a Ph.D. in systems analysis and civil engineering at Northwestern University.

HEDY E. SLADOVICH is research associate at the National Academy of Engineering Program Office. Before joining the Academy in 1988, she worked at the Marine Biological Laboratory, Ecosystems Center, in Woods Hole Massachusetts; in industry; and as a free-lance researcher in science education and policy. Ms. Sladovich's interests revolve around the interactions of technology, society, and environment. Ms. Sladovich holds a bachelor's degree in biology from Oakland University.

VICTORIA J. TSCHINKEL is senior consultant with the law firm of Landers and Parsons in Tallahassee, Florida, where she manages consultant teams to solve complex environmental problems and works with clients to resolve issues of concern to regulators and the public. From 1981 to 1987 Ms. Tschinkel served as Secretary of the Florida Department of Environmental Regulation. Under her leadership, the state's water quality standards were completely rewritten and a state water policy was adopted, including passage of a wetlands bill and substantial improvement in laws and agency programs to protect Florida's aquifers from contamination. The state also adopted a comprehensive growth management plan in 1985, and in 1986 adopted legislation to pay for cleanup of water supplies contaminated by leaking underground petroleum storage tanks. Ms. Tschinkel held positions in teaching and research prior to joining Florida's state government in 1974. She has served on numerous state and national advisory boards and committees, including appointments to the U.S. Department of Energy Research Advisory Board, the NRC Space Applications Board, the Environmental Protection Agency's Toxic Substances Advisory Committee, and the Department of Energy's Advisory Committee on Nuclear Facility Safety. Ms. Tschinkel is a zoology graduate of the University of California at Berkeley.

# Index

## A

Accidents and accident prevention
  automobiles, 51, 194
  Chernobyl, 163, 199
  nuclear facilities, 2, 104, 110–111, 163, 199–201
  Three Mile Island, 2, 163, 199
  *Valdez*, 3
Acid rain, 37, 80, 161, 169
Aerosols, 65, 142–143, 145, 168
  *see also* Chlorofluorocarbons
Agriculture, 192–193, 194
  hydroponic, 27
  irrigation, 161
Air Force, 122
Air pollutants, specific
  carbon dioxide, 10, 11, 27–28, 108–109, 111–112, 126–127, 134, 135, 196–198
  carbon monoxide, 29–30, 83
  chlorofluorocarbons, 3, 12, 18, 30–31, 65, 66, 137–154, 168–169
  fluorine, 37, 138
  freon, 168, 169
  lead, 65–66
  nitrogen oxides, 16, 31, 79, 83–84, 101, 103, 106–107, 109–110, 139, 169, 170, 172, 175, 179
  particulates, 176
  sulfur oxides, 10, 18, 31–32, 37, 97, 100–102, 106–111, 140, 168, 189
Air quality and pollution, 26
  acid rain, 37, 80, 161, 169
  chemical plants, 174
  clean coal technologies, 10, 18, 97, 106–111, 175–176
  electric generation, 100–106
  London, 79
  plastics incineration, 178, 184
  smog, 3, 169
  stratosphere, supersonic transport, 79, 139
  *see also* Chlorofluorocarbons; Combustion processes; Emission controls; Ozone and ozone layer
Air transport, 78, 116, 122–123, 194
  National Aeronautic and Space Administration, 148
  stratospheric ozone, 79, 139
Alaska, 3
Alcohol fuels, methanol, 76, 131
Alliance for Responsible CFC Policy, 143, 147, 150
Alloys, 56
Aluminum, 36–37, 56, 64
American Chemical Society, 168, 169

211

American Electric, 107–108
American Refrigeration Institute, 150
Ammonia, 45–46, 140
Ammonium chloride, 24, 168
Antarctic region, 138, 145, 148, 151
Appliances, dematerialization, 51, 52
Ardekani, Siamak A., 7–8, 50–69
Army Corps of Engineers, 161
Artificial intelligence, 111
AT&T, 85, 87
Atlantic Ocean, 66
Ausubel, Jesse H., 1–20, 50–69, 70–91
Automobiles, 189
 air pollutants, 30, 76–78, 83–84
 catalytic converters, 76, 77–78, 83, 101
 dematerialization, 51, 56–57, 59, 65–66, 85–86
 gasoline, 169
 plastics, 88
 road transport infrastructure, 73, 76
 steel consumption, 8, 56–57, 58
 tires, 16, 52, 54–55
Ayres, Robert U., 4, 6–7, 11, 17, 19, 23–49, 63, 84, 86

## B

Balzhiser, Richard E., 10–12, 95–113
Banks and banking, waste production, 61
Basins (topographic), 165
Bays (topographic), 163
Behavioral factors, *see* Human behavior; Social factors
Bhopal, 163
Biodegradability, 178, 189
Biosphere, energy system, 39–43
Biotechnology, 175, 188
 energy source, 35
 information technology and, 62–63
Byproducts, 162
 natural gas as, 118–119
 paper, 57, 61–64
 sales, 33
 *see also* Recycling; Waste residuals

## C

Cadmium, 38
Calcium carbonate, 24
California, 3, 108–109, 175, 178, 179
Canada, 146
Canals, 73

Cancer, 31, 66
Capacity, industrial, 97, 98, 100, 103–106
Capital investment
 electricity, 98, 105, 120
 energy, 130
Carbon dioxide, 135
 clean coal technologies, 10, 11, 108–109, 112
 emissions, 27–28, 126–127, 134
 Greenhouse effect, 11, 28–29, 111–112, 196–198
Carbon monoxide, 29–30, 83
Catalytic converters, 76, 77–78, 83, 101
Chemical Manufacturers Association, 138
Chemicals and chemical industry, 162, 193–194
 Alliance for Responsible CFC Policy, 143, 147, 150
 American Chemical Society, 168, 169
 atmospheric chemistry, *see* Air quality and pollution; Greenhouse effect; Ozone and ozone layer
 Bhopal, 163
 chromatography, 137–138
 Dow Chemical, 108
 Du Pont Company, 137–138, 139–140, 142, 147, 148, 150, 152
 efficiency, 32
 geochemical cycles, 37–38
 hazardous wastes, 33, 171, 173–177
 highway maintenance, 78
 history, recycling, 24–25
 incineration, 178, 183, 184, 188
 industry trends, 171
 Leblanc process, 24
 Occidental Chemical Corporation, 164–165
 plant design, 171, 173–177
 solvents, 177
 toxic waste disposal, 33
 *see also* Combustion processes
Chemicals, specific, 31
 ammonia, 45–46, 140
 ammonium chloride, 24, 168
 cadmium, 38
 calcium carbonate, 24
 chlorine, 24–25
 creosote, 72
 cryolite, 36–37
 DDT, 168, 193–194
 dichlorofluoromethane, 168

INDEX 213

HFC-134a, 150
hydrogen, 12, 16, 35, 76–77, 80–81, 82, 120, 131
hydrogen chloride, 24
insecticides and pesticides, 2, 168, 192–194
methanol, 76, 131
methyl chloride, 168
organic compounds, 177
phosphates, 131, 164–165
polyethlyene, 184
polymers, 57, 178
sodium sulfate, 24
*see also* Air pollutants, specific; Plastics
Chernobyl, 163, 199
Chesapeake Bay, 163
China, 198
Chlorine, 24–25
Chlorofluorocarbons (CFCs), 3, 12, 18, 30–31, 65, 66, 137–154, 168–169
Cholera, 160
Chromatography, 137–138
Cities, *see* Urban areas
Clean Air Act, 100–101, 145
Clean Lakes program, 165
Climate
 drought, 161, 183, 203
 *see also* Greenhouse effect
Coal and coal mining, 9, 82, 112
 clean power technologies, 10, 18, 97, 106–111, 175–176
 gasification, 11, 97–98, 106–111
 history of development, 115–116
 power plant emissions, 100–103
 sulfur oxides, 10, 18, 37, 97, 100–101, 102, 106–111
Coal tar, 24
Cogeneration, 97, 120, 131
Colorado-Ute Power Plant, 107
Combustion processes, 188
 carbon monoxide, 29–30
 clean coal technologies, 10, 18, 97, 106–111, 175–176
 coal, 10, 18, 97, 106–111
 fluidized-bed, 106–108, 109, 110
 incineration, 178, 183, 184, 188
 nitrogen oxides, 16, 31
 sulfur oxides, 31–32
 waste, 175
Communications
 infrastructure, 51

mail services, 62
telecommunications, 85, 87
telegraph, 75
Composite materials, 57, 84, 86
 electric power and, 97
Computers and computer science, 88
 dematerialization, 52–53
 electric power technology, 97, 111
 waste products, 61–62
Concrete, 57
Conservationism, 17, 72
Construction industry, 67
Consumers, 187
 consumer products, 19, 51, 52, 65, 177–178, 184–185
 in macroeconomics, 4, 6
Consumption
 chlorofluorocarbons, 142, 144
 energy, 10, 57, 60, 117
 heavy metals, 26–27, 28–31
 hydrocarbon fuels, 79–80
 in macroeconomics, 4, 6
 natural gas, 116–117
 projections, 84
 resource depletion myth, 114–116
 steel, 8, 56–57, 58
 tires, 54
 *see also* Dematerialization; Supply/demand
Coolants, 168
 chlorofluorocarbons, 140–141, 145
 reactors, 199, 200
Cornell University, 168
Cost analyses, 185
 chlorofluorocarbons, 143, 148
 Greenhouse effect, 198
 manufacturing design, 188
 natural gas, 119–122
 *see also* Prices
Court cases, *see* Litigation
Creosote, 72
Cryolite, 36–37
Cultural factors, *see* Historical perspectives; Social factors

D

DDT, 168, 193–194
Demand, *see* Supply/demand
Dematerialization, 6, 14, 50–69, 85–86
 definition, 7–9, 50–51
 *see also* Recycling

Demography, *see* Population
de Neufville, Richard, 201
Developing countries, 169
  diffusion of technology, 17–18
  green revolution, 192–193
  nitrogen oxide emissions, 172
  waste dematerialization, 8
Dichlorofluoromethane, 168
Diffusion, *see* Technological diffusion
Diseases and disorders
  cancer, 31, 66
  cholera, 160
  typhoid, 2
  *see also* Public health
DNA, 42
Dow Chemical, 108
Drought, 161, 183, 203
Du Pont Company, 137–138, 139–140, 142, 147, 148, 150, 152

## E

Economics
  banks and banking, 61
  capital investment, 98, 106, 120, 130
  chlorofluorocarbons, 140–141, 147, 148, 150, 152, 168
  common property and, 27
  energy systems, 129, 131
  Gaia hypothesis, 33–36
  incentives, economic, 17, 34–35, 65, 96
  innovation and, 32–36, 70–89, 169
  macrostructure, 4
  mass flows, 23, 25–27
  social forces, 183
  waste management, 15–16, 26
  *see also* Consumers; Cost analyses; Developing countries; Market forces; Prices
Economies of scale, 53
Education, 187
  engineering, 179–180, 201–202, 203
  multidisciplinary, 18, 179
  public, 19, 195, 200
Efficiency
  chemical processes, 32
  energy, 16–17, 32, 83–84, 96, 98, 106, 130; by country, 59
  gas turbines, 97, 119–124, 125
Electric Generation Expansion Analysis System, 122, 123–124

Electric power
  automobiles, 76
  carbon dioxide, 108–109, 111–112, 134
  clean coal technologies, 10, 18, 97, 106–111
  coal generated, 100–103
  cogeneration, 97, 120, 131
  efficiency, 106
  gas, 10–12
  historical trends, 99, 100, 104–105, 107
  hydroelectric, 103–104
  innovation, 96–97
  industrial capacity and utilization, 97, 98, 100, 103–105
  nuclear energy, 16, 35, 83, 199–201
  Power Industry and Industrial Fuel Use Act, 117
  productivity, 97, 99–100
  regulatory impact, 98, 100–106
  safety, 104, 110–111
  superconductors, 98, 171
  *see also* Power plants
Electric Power Research Institute, 96–98, 122
El-Masri, M. A., 122–123
Emission controls, 131–132
  automobiles, 30, 76–78
  carbon monoxide, 30
  catalytic converters, 76–78, 83, 101
  fluorine, 37
  hydrogen energy, 131
Energy systems, 2, 189
  alternative, 3, 9–11, 35, 97–98, 99, 116–136, 198
  biosphere, 39–43
  biotechnology, 35
  consumption trends, 82–83
  demand, evolution, 81–83
  developmental history, 81–84
  efficiency, 16–17, 32, 83–84, 96, 98, 106, 130; by country, 57, 59
  fossil fuels, 32, 114–136, 43, 197–198
  hydrocarbon fuels, 79–81, 86, 126
  hydrogen, 12, 16, 35, 76–77, 80–81, 82, 120, 131
  integrated, 130–133, 135
  petroleum and petroleum products, 3, 9, 82, 116, 118–119, 126, 135
  production, 10
  projections, 97, 116, 127–129
  resource depletion myth, 114–116

wood, 82, 115, 126
  *see also* Coal and coal mining;
    Dematerialization; Electric power;
    Natural gas; Nuclear energy
Engineering Research Center for
    Hazardous Substances, 175
Engineers and engineering
  Army Corps, 161
  design, 129, 169–174, 188, 202
  education, 179–180, 201–202, 203
  geotechnical, 3
  innovation, 9
Environmental engineering, 14, 19–20
  innovations, 16–18
Environmental Impact Statements, 185
Environmental monitoring, 16
Environmental Protection Agency, 189
  chlorofluorocarbons, 143, 145
  Clean Lakes program, 165
  environmental problems, priorities, 66
  fluorine emissions, 37
  lead emissions, 65–66
Estuaries,
  Chesapeake Bay, 163
  Everglades, 161, 162
Europe, 63, 79, 117, 120, 145, 146
Eutrophication, 163
Evans, Harold M., 184
Everglades, 161, 162
Evolutionary forces, 23
  *see also* Historical perspectives;
    Technological innovation
Expert systems, 111

## F

Federal Water Pollution Control Act of
    1972, 165
Fertilizers, 192
Fish and fisheries, 163
  mercury, 2
Flagan, Richard C., 175
Flood control, 161, 165
Florida, 160–161, 164–165
Fluidized-bed combustion, 106–108, 109, 110
Fluorine, 37, 138
  *see also* Chlorofluorocarbons
Food
  mercury poisoning, 2
  pesticides, 2
Food Service and Packaging Institute, 150

Forecasts, *see* Projections
Forests and forest products, 26
  conservation, 72
  *see also* Wood and wood products
Forest Service, 72
Fossil fuels, 32, 43
  advanced systems, 114–136
  carbon dioxide emissions, 126–127, 197–198
  *see also specific fuels*
France, 98
Freon, 168, 169
Friedlander, Sheldon K., 4, 11–12, 14, 15, 17, 167–181
Frigidaire, 168
Frosch, Robert A., 1–20
Fusion energy, 97, 171

## G

Gaia hypothesis, 33–36
Gas, *see* Natural gas
Gas chromatography, 137–138
Gasification, 11, 97–98, 106–111
Gasoline, 169
Gas turbines, 97, 119–124, 125
General Motors, 168
Geochemical cycles, 37–38
Geotechnical engineering, 3
Germany (Federal Republic), 63
Glas, Joseph P., 12, 18, 137–155
Global strategies and systems, 4, 18, 95, 96, 126, 169, 184, 185–186, 187, 188, 194, 202
  *see also* Greenhouse effect; International
    organizations and agreements;
    Ozone and ozone layer
Goeller, H. E., 86
Gray, Paul E., 2, 14, 16, 19, 192–204
Greenhouse effect, 16, 18, 80, 125–126, 163, 183, 195–198
  carbon dioxide, 11, 28–29, 111–112, 196–198
Green revolution, 192–193
Gross National Product, 99
  dematerialization and, 55, 57

## H

Haagen-Smit, Arie J., 175
Halogens, 17
  *see also* Chlorofluorocarbons

Hazardous waste
  chemical, 33, 171, 173–177
  Engineering Research Center for Hazardous Substances, 175
  heavy metals, 26–27, 28–31, 36, 161
  medical, 66–67
  process control, 179, 188
  product design, 4
Heavy metals, 26–27, 28–31, 36, 161
Herman, Robert, 1–20, 34, 50–69, 84–86
HFC-134a, 150
Hibbard, W. R., 55
Highways, *see* Road transport
Historical perspectives, 3
  chemical industry, 24
  chlorine, 24–25
  chlorofluorocarbons, 137–154, 168–169
  electricity, 99, 100, 104–105, 107
  emissions data, 37
  energy resources and systems, 79–84, 99, 100, 114–116
  environmental movement, 159–165
  evolutionary forces, 23
  hydrocarbon fuels, 79–81
  London air pollution, 79
  Midgley, Thomas, 168, 169
  natural gas, 24
  railroads, 70–73, 76, 118
  regulatory trends, 167–169
  water resource management, 159–165
  *see also* Technological innovation
Horses, transportation, 76–77
Human behavior, 3
  reactor design, 16
  *see also* Public opinion; Social factors
Hydrocarbon fuels, 79–81, 86, 126
Hydroelectric power, 103–104
Hydrogen, energy source, 12, 16, 35–36, 76–77, 80–81, 82, 120, 131
Hydrogen chloride, 24
Hydroponics, 27

# I

IBM Corporation, 52
Incentives, economic, 34–35
  energy resources, 17, 96
  recycling, 65
  tax, 17
Incineration, 178, 183, 184, 188
Industrial capacity and utilization,
  electricity, 97, 98, 100, 103–105

Industrial metabolism, 4–5, 7, 9, 11, 14, 17, 19, 23–46
  *see also* Recycling
Industrial processes
  design, 129, 169–174, 188, 202
  evolution, 32–33
  transformation, 43–46
  *see also* Dematerialization
Information technology
  computers and computer science, 52–53, 61–62, 88, 97, 111
  waste paper, 61–62
Infrastructure
  communications, 51
  energy, 11, 96, 104, 135
  sewage, 160
  transportation, 73, 75, 76, 78
Innovation, *see* Technological innovation
Insecticides, 2, 168, 192–194
Institutional factors, 12–16
Integrated gasification combined cycle, 106–110
International Institute for Applied Systems Analysis (IIASA), 116–117, 131
International organizations and agreements
  electrical generation, 96, 98
  Organization of Petroleum Exporting Countries, 197–198
  ozone layer protection, 65, 142, 143–145, 147–148, 149, 150, 152, 168
  pollution, 125
  regional strategies, 96
  regulatory actions, 17–18, 65, 142, 145–146, 168, 169; *see also* Montreal Protocol
  United Nations, 143–145, 148, 149, 168, 197–198, 199
  Vienna Convention for the Protection of the Ozone Layer, 144
  *see also* Global strategies and systems
Irradiation, wastes, 16
Irrigation, 161

# J

Japan, 63, 98, 123

# K

Kansas, 72
Kettering, Charles F., 168, 169

## L

Lake Okeechobee, 161
Lakes, 165, 192–193
Landfills, 160, 162, 178, 183, 184
  chemical wastes, 33
Law, 13, 103
  Clean Air Act, 100–101, 145
  Federal Water Pollution Control Act, 165
  liability, 4
  National Environmental Policy Act, 185
  Power Industry and Industrial Fuel Use Act, 117, 120
  Surface Water Improvement and Management Act, 165
  *see also* Litigation; Regulations
Lead
  pollution, 65–66
  scrap, 64
Leblanc process, 24
Lee, Thomas H., 3, 10–12, 63, 81, 114–136
Legal issues, *see* Law; Litigation; Regulations
Liability, 4
Licensing, reactors, 200
Light-water reactors, 199–200
Litigation, 4, 13, 161–162, 162
  chlorofluorocarbons, 145
Lovelock, J. E., 33–36, 137–138
Lynn, Walter R., 14, 19, 182–191

## M

Mail services, paper waste, 62
Maintenance and repair
  electrical utilities, 120
  road systems, 78
Manne, Alan, 133
Manufacturing
  capacity and utilization, electricity, 97, 98, 100, 103–105
  design, 129, 169–177, 188, 202
  industrial metabolism, 4–5, 7, 9, 11, 14, 17, 19, 23–46
  industrial processes, 32–33, 43–46, 129, 169–174, 188, 202
  *see also* Dematerialization; Recycling
Maritime transport, 3, 182–183
Market forces, 163, 185
  chlorofluorocarbons, 143, 150
  energy, 16
  recycling, 32–35, 178
  *see also* Consumers; Consumption
Market shares
  energy technologies, 95–96
  natural gas, 11
Marland, G., 67
Massachusetts Institute of Technology, 122, 201
Mass flows, 23, 25–27
Materials-balance principles, 36–39
  biosphere, 39–43
Materials technology, 84–89
  composite materials, 57, 84, 86, 97
  minerals replacement, 84–85
  telecommunications, 85, 87
  *see also* Dematerialization, *specific materials*
Mc Carthy, Raymond, 138
Media, 182, 195
Medical wastes, 66–67
Mercury poisoning, 2
Metabolism
  biological, 39–43
  industrial metabolism, 4–5, 7, 9, 11, 14, 17, 19, 23–46
Metals
  alloys, 56
  environmental concerns, 86
  heavy metals, 26–27, 28–31, 36, 161
  ores, 26, 31
  packaging, 52
  scrap, 64
  smelting, 36–37
  substitute materials, 84–86
  *see also specific metals*
Methane, 31, 76, 116
  *see also* Natural gas
Methanol, 76, 131
Methyl chloride, 168
Midgley, Thomas, 168, 169
Military applications, 122–123
Mineral resources, 84, 164
  mass flow and, 25–26
  *see also specific minerals and metals*
Miniaturization, 52
Minnesota, 107
*Mobro,* 182–183
Models, 188
  biological, 39–43; 162
  electric power techniques, 97
  Gaia hypothesis, 33–36

industrial transformation processes, 43–46
  Montreal Protocol as, 169
  technological development, 72–73, 80–81
  *see also* Industrial metabolism; Projections
Molina, M., 138
Montana-Dakota Power Plant, 107
Montreal Protocol on Substances That Deplete the Ozone Layer, 148, 149, 150, 152, 168, 169–170
Municipalities, *see* Urban areas

## N

National Aeronautics and Space Administration, 148
National Environmental Policy Act, 185
National Research Council, 78, 184, 187
Natural gas, 126, 198
  advanced systems, 10–12, 112, 116–124
  consumption, 83
  economics, 97, 106, 109
  history, 24, 80–81
  production, 121
  turbines, 97, 119–124, 125
Natural Resource Defense Council, 145
Netherlands, 63
New Jersey, 66
News media, 182, 195
Newspapers
  municipal waste, 57–58, 61, 63
  recycling, 178
New York, 66, 178, 182–183
Nitrogen oxides, 16, 31, 79, 83–84, 101, 103, 106–107, 109–110, 169, 170, 172, 175, 179
  ozone destruction, 139
Non-point-source pollution, 163
Nonrenewable resources, 86, 89
  myths, 114–116
Northern States Power, 107, 109
Norway, 120
Nuclear energy, 83, 98, 120, 199–201
  Chernobyl, 163, 199
  coolants, 199, 200
  electric propulsion, 35
  fusion energy, 97, 171
  innovation, 16, 104, 110–111, 131
  public opinion, 200
  safety, 104, 110–111, 199–201
  Three Mile Island, 2, 163, 199

## O

Occidental Chemical Corporation, 164
Office of Technology Assessment, 175
Oil, *see* Petroleum and petroleum products
Oil pollution, 3
Organic compounds, 177
Organization of Petroleum Exporting Countries, 197–198
Ozone and ozone layer, 3, 12, 18, 30–31, 65, 66, 137–154, 163, 168–169
Ozone Trends Panel, 148, 150

## P

Packaging
  computers, 52
  Food Service and Packaging Institute, 150
  plastic, 52, 184
Panel for Alternative Fluorocarbon Toxicity Testing, 148
Paper and paper products, 86
  newspapers, 57–58, 61, 63, 178
  photocopying, 61, 62
  waste, 57, 61–64, 88
Particulates, 176
Paul, B., 63
Pennsylvania, Three Mile Island, 2, 163, 199
Pesticides, 2, 168, 192–194
Petroleum and petroleum products, 9, 82, 126
  carbon dioxide used to recover, 135
  development, 116, 118–119
  oil pollution, 3
  paper manufacturing, 62
Phosphates, 131, 164–165
Photocopying, 61, 62
Physicians, 160
Plastics, 86, 177–178
  automotive, 88
  composites, 56
  metal substitutes, 84–86
  packaging, 52, 184
  polymers, 57, 178
  waste, 61
Politics, 35, 183, 189, 201
  chlorofluorocarbons, 147–148, 198
  nuclear technology, 200
  water resources management, 163
Policy, 189, 195, 201, 203

chlorofluorocarbons, corporate policy, 139–140, 143, 147
  industrial metabolism, 7, 14
Polyethylene, 184
Polymers, 57, 178
Population
  growth, 2, 169, 173, 194
  spatial dispersion, 8, 54
Postal services, paper waste, 62
Power Industry and Industrial Fuel Use Act, 117
Power plants
  near-term problems, 95–113
  rate structure, 120
Predictions, see Projections
Preservatives, railroad crossties, 72–73
Prices
  fuel, 35, 103, 109, 197–198
  utility rate structure, 120
Production and productivity
  chlorofluorocarbons, 144
  economies of scale, 53
  electricity, 97, 98, 99, 100, 103–105
  energy, 10, 121
  Gross National Product, 55, 57, 99
  in macroeconomics, 4, 6
  tires, 54
  see also Efficiency; Supply/demand
Projections, 194
  demand, 84, 97
  energy, 97, 116, 127–129
  ozone layer/CFCs, 138–139, 140, 141, 143, 145, 146–147, 153
  reactor safety, 200–201
  resource depletion myth, 114–116
Propellants, aerosol, 65, 142–143, 145
Public health
  electric and magnetic fields, 98
  water supply chlorination, 2
Public opinion, 1–3, 163–164, 182–184, 194–195, 200

## Q

Quality control and improvement
  dematerialization and, 52–53
  energy systems, 129

## R

Radioactive pollution, nuclear reactor design, 16, 199–201

Railroads, 70–73, 76, 118
Rain, 163
  acid rain, 37, 80, 161, 169
  storm water, 160–161
Reactors, see Nuclear energy
Recycling, 4, 7, 15, 36, 162, 189
  biosphere and industrial economy, 23
  chemical industry, 24–25, 32–33, 174, 177
  cost incentives, 65
  dematerialization, 6, 7–9, 14, 50–69
  market forces, 32–35
  newspapers, 178
  paper, 63–64
  plastics, 178
  regulation vs, 164
  scrap metal, 64
Refrigeration, 140–141, 145, 150, 168
Regional strategies, 96, 122, 123–124, 186
Regulations, 10, 13, 14, 33, 163–165, 174, 175, 185, 195
  carbon monoxide emissions, 30
  chlorofluorocarbons, 65, 143–145, 147–148, 149, 150, 152, 168, 169–170
  electric power, 98, 100–106, 121–122
  international, 17–18, 65, 142, 145–146, 168, 169
  lead, 65–66
  plastics, 178
  reactor licensing, 200
  trends, 167–169
  water management, 160, 162, 163
  see also Environmental Protection Agency
Renewable resources, mass flow, 25–26
  see also Recycling
Repairs, see Maintenance and repair
Research and development
  biological models, 162
  chlorofluorocarbons, 138, 142, 148, 150, 151
  highways, 78
  plant wastes, 174–177
  waste technology, 187–189
  see also Technological innovation
Risk assessments, see Projections
Rivers, 164–165, 192–193
Road transport, 73, 76, 78
Rowland, F. S., 138
Rupp, W. H., 174

## S

Safety, nuclear, 104, 110–111, 199–201
Santa Barbara Channel, 3
Scrap metal, 64
Sensor technologies, 110–111, 188
Sewage, 160
Shapiro, Irving, 140
Ships and shipping, 3, 182–183
Shoes, 53
Silicon wafers, 52
Sludge, 66
Smelting
  aluminum, 36–37
  sulfur oxide emissions, 37
Smog, 3, 169
Social factors, 12–16, 18–19, 125, 163, 164, 185–186, 201–202, 203
  chlorofluorocarbons, 148
  consumers, 4, 6, 19, 51, 52, 65, 177–178, 184–185, 187
  dematerialization, 53–55
  education, 18, 19, 179–180, 187, 195
  innovation and, 20, 129, 193
  institutional factors, 12–16
  market forces, 16, 143, 150, 163, 178, 185
  media, 182, 195
  paper waste and, 61
  politics, 35, 147–148, 163, 183, 189, 198, 200, 201
  policy, 7, 14, 139–140, 143, 147, 189, 195, 201, 203
  population, 2, 8, 54, 169, 173, 194
  public health, 2, 98
  public opinion, 1–3, 163–164, 182–184, 194–195, 200
  resource depletion myth, 114–116
  sociotechnical systems, 72–74, 98–99, 185–187
  *see also* Historical perspectives; Law; Regulations
Sodium sulfate, 24
Solar radiation, 30, 39, 138
Solvents, 177
Southern California Edison, 107–108, 175
South Florida Flood Control Project, 161
Soviet Union, Chernobyl, 163, 199
Stanford University, 131
Statutes, *see* Law; Litigation; Regulations
Steel
  consumption, 8, 56–57, 58
  energy requirements, 130

St. Johns River, 165
Storm water, 160–161
Strategic Highway Research Program, 78
Stratosphere
  supersonic transport, 79, 139
  *see also* Ozone and ozone layer
Stratospheric Ozone Protection Plan (EPA), 145
Sulfur oxides, 31–32, 37, 189
  chlorofluorocarbons as alternative, 140, 168
  clean coal technologies, 10, 18, 97, 106–111
  scrubbers, 100–101, 102, 108
Superconductors, 98, 171
Supersonic transport, 79, 139
Supply/demand
  chlorofluorocarbons, 146
  "demandite" concept, 86, 88–89
  dematerialization and, 53–55
  electricity, 97, 98, 100, 103–105
  energy, 10, 81–83, 97, 103
  equilibrium, 31–32
  hydrocarbon fuels, 126–127
  projections, 84
  wood industry history, 72
Surface Water Improvement and Management Act, 165
Suwannee River, 164–165
Sweden, 135

## T

Taxes, subsidies, 17
Technological diffusion, 81, 88
  to developing countries, 17–18
Technological innovation, 9–10, 70–90, 189, 192–204
  aircraft, 118
  dematerialization and, 54
  electric power, 96–97, 171
  energy, 81–84, 129–130, 171
  environmental quality, 16–18, 20
  industrial processes, 32–36
  natural gas, advanced systems, 10–12, 112, 116–124
  social factors, 20, 129, 194–195
  transportation systems, 2, 70–79
Telecommunications, 85, 87
Telegraph, 75
Tenner, E., 61

Tennessee Valley Authority (TVA), 107–108
Three Mile Island, 2, 163, 199
Time series analysis, 129
Tires, 16, 52, 54–55
Toxic waste, *see* Hazardous waste
Transportation systems, 2, 70–79
  air transport, 78, 79, 116, 122–123, 139, 148, 194
  canals, 73
  energy, 82
  horses, 76–77
  infrastructure, 73
  maritime, 3, 182–183
  railroad, 70–73, 75, 76, 118
  road, 73, 76, 78
  waste, 7, 182–183
  *see also* Automobiles
Troposphere, 79
Tschinkel, Victoria J., 3–4, 12–13, 18, 19, 159–166

## U

Ultraviolet radiation, ozone layer depletion, 30–31, 138
United Kingdom, 79, 145
United Nations Environment Program, 143–145, 148, 149, 168, 197–198, 199
Urban areas
  air pollution, 29, 76–77
  sludge, 66
  solid waste, 8, 57, 60–61, 177–178, 182–183
  water supply, 161

## V

*Valdez*, 3
Vienna Convention for the Protection of the Ozone Layer, 144

## W

Waste residuals, 7, 15, 13, 23, 25–33
  construction industry, 67
  lead, 64, 65–66
  materials-balance principles, 36
  medical, 66–67
  municipal, 8, 57, 60–61, 66
  packaging, 52, 184
  paper, 57–58, 61–64, 86, 88, 178
  silicon wafers, 52
  throwaway products, 26, 65, 66
  *see also* Dematerialization; Recycling
Waste management, 14–16, 162, 187–189
  biodegradability, 178, 189
  chemical plant design, 174–177
  landfills, 33, 160, 162, 178, 183, 184
  *Mobro*, 182–183
  water, 160–161, 163
  *see also* Hazardous waste
Waterfowl, 161
Water quality and pollution
  acid rain, 37, 80, 161, 169
  Atlantic Ocean, 66
  chlorination, 2
  eutrophication, 163
  Federal Water Pollution Control Act, 165
  fertilizers, 192
  heavy metals, 26–27, 28–31, 36, 161
  insecticides and pesticides, 2, 168, 192–194
  lakes, 161, 165, 192–193
  mercury poisoning, 2
  non-point-source pollution, 163
  oil pollution, 3
  phosphates, 131, 164–165
  rivers, 164–165, 192–193
  road ice removal, 78
  sewage, 160
Water resource management, 159–165
  fish and fisheries, 2, 163
  flood control, 161, 165
Watersheds, 165
Weinberg, A. M., 67, 90, 186
Wildlife, 161
Wood and wood products
  energy source, 82, 115, 126
  railroad industry development, 70–72, 76
  recycling, 34–35
  *see also* Forests and forest products; Paper and paper products

## X

Xerox, 62

## Z

Zinc scrap, 64

$V m^3 \dfrac{t}{r} \times 1.5 \longrightarrow \, : 94.6$

1) volume = (50 m³) —— $\dfrac{50 x}{5000}$ : at

   wt (kg) : 60

$\dfrac{0.6 \times 0.6 \times 0.8 =}{\underset{2.88}{\underset{8}{36}}}$  $\dfrac{2.3 m^3 \times }{5000} \; = $